# IEIA - H-SCADA Bio-Energy Field Professional Study Guide for Examination

TERRA·MARE·ET·AER
DEFENDO

Founded in 2001

**Dr. Hildegarde Staninger, Ph.D., RIET-1,
Melinda Kidder, B.S., P.I., CESCO
Dr. Daniel F. Farrier, MD, CIET**

# IEIA
## H-SCADA Bio-Energy Field Professional Study Guide for Examination

### International Environmental Intelligence Agency (IEIA)
1770 Algonquin Trail
Maitland, FL 32751

*Administration of Certification Programs are
by Integrative Health Systems®, LLC*
Phone: 323-466-2599 or 407-628-1303   Fax: 323-466-2774
E-mail: ihs-drhildy@sbcglobal.net
To contact: IEIA's Executive Director/Inspector General
Dr. Hildegarde Staninger, RIET-1
Phone: 323-466-2599 or 407-628-1303

#### Authors
Hildegarde Staninger, Ph.D., RIET-1
Melinda Kidder, B.S., PI, CESCO
Daniel Farrier, MD, CIET

### IEIA Inspector General's Council on SCADA, Members
Inspector General, IEIA: Dr. Hildegarde Staninger®, RIET-1
Melinda Kidder, B.S., PI, CESCO
Benjamin Colodzin, Ph.D.

© September 16, 2015

Copyright © 2015 by Dr. Hildegarde Staninger, Ph.D., RIET-1
Melinda Kidder, BS, PI, CESCO
Dr. Daniel Farrier, MD, CIET

*IEIA - H-SCADA Bio-Energy Field Professional Study Guide for Examination*
by Dr. Hildegarde Staninger, Ph.D., RIET-1
Melinda Kidder, B.S., PI, CESCO
Dr. Daniel Farrier, MD, CIET

Printed in the United States of America.

ISBN 9781498431729

All rights reserved solely by the author. The author guarantees all contents are original and do not infringe upon the legal rights of any other person or work. No part of this book may be reproduced in any form without the permission of the author. The views expressed in this book are not necessarily those of the publisher.

International Environmental Intelligence Agency (IEIA)
1770 Algonquin Trail
Maitland, FL 32751
Phone: 323-466-2599 (direct to Inspector General) Fax: 327-466-2774

www.xulonpress.com

# IEIA - H-SCADA Certification Series

## An Introduction to the H-SCADA Bio-Energy Field Professional and What this Professional Means to the World

The International Environmental Intelligence Agency (IEIA) is a private State of Florida Not-for-Profit Corporation. It is a non-governmental environmental intelligence, accrediting organization, educational fellowship/scholars research. IEIA is recognized by organizations, industry, military and government for the development of the H-SCADA Methodology, as originally developed by Melinda Kidder, B.S., PI, CESCO for the assessment of the human biological system's exposure to various bio-energy fields and nanotechnology as described by SCADA disciplines.

In 2003, IEIA's Executive Director and Inspector General recognized there were no environmental, safety and health regulations specifically to address the public health concerns for any potential exposure to advanced smart materials as associated with nano and bio-technology. And in the fall of 2013 IEIA further committed their concerns through their professional associates and fellows/scholars on the use of these materials into the integration of Geo-Engineering, Biological Monitoring of Biological Systems and Environmental Monitoring that would use remote sensing networks with applications of SCADA (Supervisory Control and Data Acquisition). And in reviewing these concerns, it was unanimously recognized that professionals need to be trained in the recognition, investigation and assessment of exposure to advanced materials and the various energy fields associated with specific innovative technology.

IEIA, over the years, has recognized that there is a professional need for highly skilled and trained professionals who have gone through the extra steps of certification to become the "Elite" of their specialized expertise field. Any individual who has been certified by IEIA has been recognized for their professional integrity and dedication in defending the land, air and sea of this beautiful planet called Earth. These professionals come from all walks of life and professions and are dedicated to the truth and honesty of their professional quality of work. As an IEIA Certified Professional, they will have professional liability and general liability insurance coverage as an added benefit to their professional work and clients' concerns. The IEIA H-SCADA Bio-Energy Field Professional certification credential is recognized by Energy Medicine Professional Insurance and other carriers, who are leaders in the professional insurance needs of professionals who practice in this highly specialized field of science, healthcare and bio-engineering.

The term SCADA (Supervisory Control and Data Acquisition) is defined as a system operating with coded signals over communication channels so as to provide control of remote equipment (using typically one communication channel per remote station). SCADA applications have now grown into the health care and collaborative research arenas for its use in the applications of assessments of the Human Biological System of its own bio-energy field and the integration of sensory technology, which is called H-SCADA. The toxicological pathological concerns are great in this new arena and that is why IEIA is committed to developing professional certification programs and criteria documents to assist humanity for its own protection and well-being.

The various disciplines that will use a SCADA system are from all sources of industry, medicine, pharmaceutical, transportation, military, academia, communications, environmental, government and/or interplanetary space communications for space medicine, defense and commercialization. Due to the multi-faceted aspects of any SCADA auspices it should be realized there will be continuous applied integration of multiple networks and systems, especially with the development of electromagnetic, holographic gradient, high Wi-Fi and Li-Wi systems.

A survey was conducted through IEIA's Inspector General's Council and its alliance partners as well as other private industrial and academic fellows/scholars for the joint supported efforts in developing a Certification Program for H-SCADA applications through the through IEIA. The H-SCADA Bio-Energy Field Professional Study Guide for Certification and Examination was developed by the Fellows and Scholars of IEIA, as listed below:

Melinda Kidder, B.S., PI, CESCO; Dr. Daniel Farrier, MD, CIET; and the Executive Director/Inspector General of IEIA, Dr. Hildegarde Staninger, RIET-1.

The IEIA Inspector General's Council on SCADA and its related innovative technology was established in September 2015. It was established to address the basic criteria to be included in a special study guide to be used for advanced professional training and development. Specifically it would address professionals' and future professionals' needs to seek knowledge in understanding SCADA, as it applies to the environment, science, medicine, neurosciences/mental health, law enforcement, public health, engineering and other applied professional fields. This special training would then allow individuals and professionals to be able to apply for the IEIA H-SCADA Bio-Energy Field Professional Certification Examination process. This would continue their thirst for Professional Developmental Training for advanced H-SCADA and other network monitoring program training, as well as their applied field experience to augment their professional expertise. These areas would include the field analysis of innovative advanced smart materials, nanotechnology, bio-life systems, hardware/software equipment, hand-held monitoring equipment and other related technologies.

The term SCADA (Supervisory Control and Data Acquisition) was originally a type of Industrial Control System (ICS). It has now branched out into medicine, pharmaceuticals, agriculture, academia, military and many other remote-controlled computer systems. These systems consist of data for many other remote-controlled computer systems of data storage, analysis and real-time monitoring of specimens. The aspects of Industrial Control Systems are computer-controlled systems that monitor and control industrial processes, which exist in the physical world. SCADA systems historically distinguish themselves from other ICS systems by being large scale processes that can include multiple sites and long distances (Example: Earth, Satellite, Space and Cosmos). These processes include industrial, infrastructure and facility-based processes.

- Industrial processes include those of manufacturing, production, power generation, fabrication, and refining. Many run in continuous batch, repetitive and/or discrete modes.

- Infrastructure processes may be public or private, and include water treatment and distribution, wastewater collection and treatment, oil and gas pipelines, electrical power transmission and distribution, wind farms, civil defense siren systems, and large communication systems, as well as, various types of recording and archiving data bases.

- Facility processes occur both in public facilities and private ones, including buildings, airports, hospitals, schools, ships and space stations. They monitor and control heating, ventilation, and air conditioning systems (HVAC) as well as, access and energy consumption. In medicine, one may monitor data gathered from a human body remotely or from a laboratory experiment.

Recent developments in nanotechnology, bioengineering, neuroscience, neuro-telepathy, biomedicine and life sciences have now incorporated nano-silicon-CMOS, CMOS, MOSFET and MITRI Advanced Computer Systems into integrated advanced materials software and data storage/monitoring. The use of META surface materials, infrared glass, bio-glass and/or sol-gel based mid-infrared evanescent wave sensors in nanotechnology will allow for holographic gradient Wi-Fi systems. Further applications in the very near future under real-time monitoring networks for health, biometrics and other aspects as applied to Human-Machine-Computer Interface approaches to custom CLOUD Architectures will be part of the H-SCADA systems. IEIA's professionals look forward to identifying these piece-part tools and systems under H-SCADA Methodology through Advanced Training to validate exposures, toxicological risk factors and remediation methodology.

It is further hoped by IEIA that young scientists, engineers and other professionals will apply their desire to further their research in H-SCADA Bio-Energy Field studies as the technology develops further into the global network and its mitigation.

Dr. Hildegarde Staninger®, RIET-1
Executive Director & Inspector General
INTERNATIONAL ENVIRONMENTAL INTELLIGENCE AGENCY (IEIA)
September 17, 2015

## A Very Special Note Of Appreciation
September 17, 2015

The International Environmental Intelligence Agency (IEIA) and its Fellows and Scholar Members want to instill a special THANK YOU to Ms. Shirley "Shoshanna" Allison for all of the help, research, edits, formatting, assistance and support she gave us over the last year of this project. Your patience and professionalism was greatly appreciated by all of us through the many revisions of the Study Guide.

Another special appreciation Thank You goes to Dr. Benjamin Colodzin, Psychologist, for being the first one to review and apply the IEIA H-SCADA Bio-Energy Field Professional Study Guide and Required Examination document through all appropriate testing applications.

We want to give a special thank you to Roberto Mentuccia, B.S., CESCO, who developed the IEIA Certification Brochure, "What is H-SCADA Bio-Energy Field Professional Certification?", as well as his technical contributions to the Study Guide for agriculture food safety and security concerns as they relate to nanotechnology and bio-engineering applications.

IEIA INSPECTOR GENERAL COUNCIL
September 17, 2015

# IEIA - H-SCADA Certification Series

## IEIA - H-SCADA Bio-Energy Field Professional
## Study Guide for Examination

### TABLE OF CONTENTS

**H-SCADA Certification Exam - Required Study Guide - -** *Cover Sheet…..5*

*An Introduction to H-SCADA and What It Means to the World……………….7 thru 10*

*Acknowledgements………………………………………………………………………………….11*

*Table of Contents ………………………………………………………………… 13 thru 15*

*Important Note ……………………………………………………………………………………..16*

*Study Guide Overview – TEXTS …………………………………………………………..17, 18*

*Study Guide Overview – ARTICLES …………………………………………………….18, 19*

*Study Guide Overview …………………………………………………………………….19, 21*

*Outline of Topics to be Covered………………………………………………………….20, 21*

**COURSE OVERVIEW……………………………………………………………………….22 thru 31**

**MODULE 1** – History of SCADA Systems and Common System Components..
…………….32 thru 51
*Lesson 1: NCS TIB 04-1: National Communications System (NCS)……….. 33*
*Lesson 2: TIB Analysis, Observations and Conclusions……………………………..39*
*Lesson 3: NCS role in SCADA systems, Utilities, IEEE and Monitoring……..47*

**MODULE 2** – Critical Infrastructure: Homeland Security and Emergency
Preparedness (by Robert Radvanovsky)        52 thru 88
*Lesson 1: Chapter 1- Introduction to Critical Infrastructure Preparedness..54*
*Lesson 2: Chapter 2 – Regulations and Legislation…………………………………..58*
*Lesson 3: Chapters 3, 4 and 5 –(pgs 41-118)……………………………………...63*
*Lesson 4: Chapter 6 (pgs 119-144) Chapter 7 (pgs 145-173) and
 Chapter 8 (pgs 175-198)………………………………………………..……….70*
*Lesson 5: Chapter 9 (pgs 199-237), Chapter 10 (pgs 239-259) and
 Chapter 11 (pgs 261-287)………………………………………………………..80*

**MODULE 3** – Advances In Computer: Volume 71 Nanotechnology, Edited by
Marvin V. Zelkowitz ..................................................89 thru 112
Lesson 1: Chapter 1 (pgs 1-39)................................................90
Lesson 2: Nanobiotechnology: An Engineer's Foray into Biology -
Chapter 2 (pgs 39-102) ................................................97
Lesson 3: Toward Nanometer –Scale Sensing Systems –
Chapter 3 (pgs 103-166).............................................103
Lesson 4: Simulation of Nanoscale Electronic Systems –
Chapter 4 (pgs 167-251)..............................................107
Lesson 5: Identifying Nanotechnology in Society –
Chapter 5 (pgs 251-271)..............................................109
Lesson 6: Chapter 6 (pgs 273-296)..........................................111

**MODULE 4 -** Design for Manufacturing and Yield for Nano-Scale CMOS
by Charles C. Chiang and Jamil Kawa ............................ 113 thru 121
Lesson 1: Chapters 1, 2, 3 and 4 (pgs 1-168) and Preface
(pgs XX1-XXV).............................................................113
Lesson 2: Chapters 5, 6, 7 & 8 (pgs 99 – 242)............................121

**MODULE 5:** FCC Online Table of Frequency Allocations
(47 C.F.R. Statute 2.106) Revised May 25, 2012     122 thru 124
Lesson 1: Federal Register (pgs 1-97)........................................123
Lesson 2: FCC Document (pgs 68-168)......................................124

**MODULE 6:** FDA/OSEL Annual Report 2007 by Food and Drug Administration
Center for Devices and Radiological Health     125 thru 126
Lesson 1 and Lesson 2: Office of Science and Engineering Laboratories
2007 Highlights for Laboratory and their augmentation
through OSEL (pgs 1-73)..............................................126

**MODULE 7:** Guidelines on the Evaluation of Vector Network Analyzers (VNA)
Euramet     127 thru 130
Lesson 1 and Lesson 2: Read pgs (1-19)......................................124

**MODULE 8:** Telemedicine and e-Health: Abstracts from the American
Telemetric Association     1  thru 133
Lesson 1 and Lesson 2:  Telemedicine...........................................131

**MODULE 9:** Energy Field, Radiation, Health and Safety Issue     134 thru 138
Lesson:     Read pgs (1-9) Scan Tech...................................................135

# IEIA - H-SCADA Certification Series

**MODULE 10:** SANDIA Report: SAND 2004 – 5625 Unlimited Release
                                                                139 thru 141

Lesson: SANDIA Report: SAND 2004 – 5625 Unlimited Release –
Read (pgs 1-78)..................................................................................140

**MODULE 11:** Nano Air Vehicles A Technology Forecast    142 thru 148
Lesson: Read (pgs 1-35)..................................................................143

**MODULE 12:** Technical Supplementary Information and Articles  149 thru 151
Lesson: Read specific Articles Listed on page 137 above......................150

**APPENDIX A – Acronyms**................................................ 152 thru 154

*Appendix B – Reference Texts, Articles with hot-links for e-Book users.*
                                                                  .................155 thru 158

**BROCHURE – IEIA – H-SCADA CERTIFICATION**.......................... 159 thru 160

**ENCOUNTERING CLIENTS CLAIMING EXTERNALLY CONTROLLED "IMPLANTS" ARE AFFECTING THEIR HEALTH: An advisory for healthcare professionals....**Ben Colodzin, Ph.D........................161 thru 163

**H-SCADA PROTOCOL EVALUATION METHODS and EQUIPMENT...**
                                                                         ..........165 thru 167

*IEIA Logo*.................................................................................169

*Answers to Study Guide Questions*..........................................171 thru 179

**NOTES** ................................................................................180 thru 184

## IEIA - H-SCADA Bio-Energy Field Professional
## Study Guide for Examination

*Important Note:*

Occasionally books are revised or removed from publication and the corresponding study guide must be updated. You should compare the ISBN (International Standard Book Number) on each book you are referenced to use or receive to the ISBN in the corresponding study guide. This number is an identifier that is updated whenever a book is released in a new edition. Because we update our study guides and certification exams to reflect changes in our textbooks, using an older edition of the book with a newer version of the exam may result in low test scores. If the ISBN on your study guide does not match the ISBN on the book, please call IEIA's administrative office.

You can find the ISBN on the outside back cover of the textbook and also on the page following the title page with information concerning the copyright.

Due to the ever-growing changes in Supervisory Control and Data Acquisition (SCADA) Systems implementation into industrial processes, software, computer, facility processes and developments and interfacing of nanotechnology, biomedicine and life sciences will be addressed in the certification exam through state of the art professional papers on the subject matter when text books are not available. IEIA has established the SCADA REFERENCE LIBRARY on its official website for each of their certification exams with required text and professional peer-reviewed articles to assist examinees in preparing for their certification exam.

# IEIA - H-SCADA Bio-Energy Field Professional
# Study Guide for Examination

**TEXTS:**

### NCS TIB 04-1: National Communications System
Technical Information Bulletin 04-1
Supervisory Control and Data Acquisition (SCADA) Systems
Office of the Manager, National Communications System, P.O. Box 4052
Arlington, VA 22204-4052 © October 2004

### Critical Infrastructure: Homeland Security and Emergency Preparedness
Robert Radvanovsky, ISBN: 0-84593-7398-0
CRC Press/Taylor & Francis (A CRC Press Book), Boca Raton, FL © 2006

### Advances in Computers: Volume 71 Nanotechnology
Edited by: Marvin V. Zelkowitz, ISBN: 978-0-12-373746-5
Elsevier/Academic Press, New York, New York © 2007

### Design for Manufacturability and Yield for Nano-Scale CMOS
Charles C. Chiang and Jamil Kawa, ISBN: 978-1-4020-5187-6
Springer, Netherlands © 2007

### FCC Online Table of Frequency Allocations (47 C.F.R. Statue 2.106)
Federal Communications Commission Office of Engineering and Technology
Policy and Rules Division, Revised on May 25, 2012
US Governmental Printing Office, Washington, DC

### FDA/OSEL Annual Report 2007
Food and Drug Administration's Center for Devices and Radiological Health (CDRH) and Office of Science and Engineering Laboratories (OSEL)
Larry G. Kessler, Sc.D., Director OSEL/DCRH & Chair, Global Harmonization Task Force

**Guidelines on the Evaluation of Vector Network Analyzers (VNA)**
**EURAMET (European Association of National Metrology Institutes)**
EURAMET cg-2, Version 2.0 (03/2011) Previously EA-10/12
Calibration Guide, Braunschweig, Germany  e-mail:
secretariat@euramet.org   Phone:   +49 531 592 1960

**Telemedicine and e-Health:  Abstracts from the American Telemedicine Association**
Eighteenth Annual International Meeting and Exposition,
May 5-7, 2013 – Austin, Texas  ISSN 1530-5627  Mary Ann Liebert, Inc. Publishers

**Section 8:  Health Effects Associated with Exposure of Industrial Workers to Radiofrequency Waves**
RF Toolkit-BCCDC/NCCEH pages 191 – 216, CDC, Atlanta, GA

**Sandia Report:  SAND2004-5625 Unlimited Release Printed November  2004:** Microfabrication with Femtosecond Laser Processing – (A) Laser Ablation of Ferrous Alloys, (B) Direct-Write Embedded Optical Waveguides and Integrative Optics in Bulk Glass.
Pin Yang, George R. Burns, Jeremy A. Palmer, Marc F. Harris, Karen L. McDaniel, Junpeng Guo, G. Allen Vawter, David R. Tallant, Michelle L. Griffith and Ting Shan Luk
Sandia National Laboratories, Albuquerque, New Mexico, 87185 and Livermore, California

**Nano Air Vehicles  A Technology Forecast**
William A. Davis, Major, USAF, April 2007
Blue Horizons Paper, Center for Strategy and Technology, Air War College.  In accordance with Air Force Instruction 51-303, it is not copyrighted, but is the property of the United states government.

**ARTICLES**:

***Extremely Scaled Silicon Nano-CMOS Devices***
Leland Chang, Yang-Kyu Choi, Daewon Ha, Pushkar  Ranade, Shiying Xiong, Jeffrey Boker, Chenming Hu and Tsu-Jae King
Proceedings of the IEEE, Vol. 91, No. 11, November 2003

***Designs for Ultra-Tiny, Special-Purpose Nanoelectronic Circuits***
Shamik Das, Alexander J. Gates, Hassen A. Abdu, Garrett S. Rose, Carl A. Picconatto and James C. Ellenbogen IEEE Transactions on Circuits and Systems – I: Regular Papers, Vol. 54, No. 11, November 2007

***General Recipe for Flatbands in Photonic Crystal Waveguides***
Omer Khayam and Henri Benisty
17 August 2009/Vol. 17, No. 17/ Optics Express 14634

## STUDY GUIDE OVERVIEW

The topics to be covered for the SCADA Certification Exam Study Guide are primarily composed of the history, technology, development, implementation, integrated computer hardware, advanced software applications and other smart and/or nanotechnology integration into various advanced computer systems analysis networks and Cloud technologies.

SCADA (Supervisory Control and Data Acquisition) Systems was originally a type of Industrial Control System (ICS) that has now branched out into medicine, pharmaceuticals, agriculture, academia, military and many other remote-controlled computer systems of data storage, analysis and monitoring. The aspects of Industrial Control Systems are computer-controlled systems that monitor and control industrial processes which exist in the physical world. SCADA systems historically distinguish themselves from other ICS systems by being large-scale processes that can include multiple sites, and long distances (Example: Earth, Satellite, Space and Cosmos). These processes include industrial, infrastructure and facility-based processes as described below:

- Industrial processes include those of manufacturing, production, power generation, fabrication, and refining, and many run in continuous, batch, repetitive, or discrete modes.
- Infrastructure processes may be public or private, and include water treatment and distribution, wastewater collection and treatment, oil and gas pipelines, electrical power transmission and distribution, wind farms, civil defense siren systems, and large communication systems.
- Facility processes occur both in public facilities and private ones, including buildings, airports, hospitals, schools, ships and space stations. They monitor and control heating, ventilation and air conditioning systems (HVAC), access and energy consumption. In medicine they may monitor data gathered from a human body.

- Recent developments in nanotechnology, biomedicine and life sciences have now incorporated Nano-CMOS, Moffett and MITRI Advanced Computer Systems into an integrated platform of advanced materials, software and data storage/monitoring.

**Outline of Topics to be Covered for Certification Exam through this Study Guide:**

1. History of SCADA

2. Common System Components

3. Systems Concepts (Nano, Micro and WI FI)

4. Human-Machine Interface

5. Hardware Solutions (Supervisory station, Operational philosophy, and Communication infrastructure and methods

6. SCADA architectures: First generation (Monolithic); Second generation (Distributed), Third generation (Networked) and Fourth generation (Cloud plus)

7. Security and Public Privacy Issues

8. SCADA Protocols

9. Deploying SCADA Systems (Twisted-pair metallic cable, Coaxial metallic cable, fiber-optic cable, power line carrier, satellites, leased telephone lines, very high frequency radio (Terahertz); Ultra-high frequency radio (point-to-point, multiple address radio systems, spread spectrum radio, microwave radio, Terahertz).

10. Security vulnerability of SCADA Systems (Attacks against SCADA systems and Developing a SCADA Security Strategy)

11. SCADA Standards Organizations
The Institute Of Electrical and Electronics Engineers (IEEE); American National Standards Institute (ANSI); Electric Power Research Institute (EPRI), International Electro-technical Commission and DNP3 Users Group

12. SCADA Governmental and International Interagency Organizations National Communications Systems and Federal Telecommunications Standards Program for Academia, Military, Agency and Industrial Overlapping Technologies.

13. Observations and Conclusions

14. Current and Future Recommendations

Appendix A- Acronyms

Appendix B – Reference Texts, Articles with hot-links for e-Book users.

# IEIA - H-SCADA Certification Series

**MODULE 1: History of SCADA Systems and Common System Components**

**Lesson 1**: *NCS TIB 04-1: National Communications System (NCS)*
- Introducing the concepts of SCADA systems.
- Identifying what components make up a typical SCADA system.
- Plotting the evolution of SCADA systems through its monolithic, distributed, and networked evolution
- Looking at the ways in which a SCADA system can be deployed.
- Examining the protocols used in these systems currently, as well as the standards and potential future SCADA protocols.

**Lesson 2**: *TIB Analysis, Observations and Conclusions*
- SCADA systems began in 1960's and have evolved as technology changes.
- SCADA systems' evolutionary role with mainframe-based to client/server architectures. Common communications protocols like Ethernet, Internet, and TCP/IP to transmit data from the field to the master control unit.
- SCADA protocols evolved from closed proprietary systems, to open system allowing designers to choose equipment and monitor unique systems from mixed vendor equipment.

**Lesson 3**: *NCS role in SCADA systems, utilities, IEEE and Monitoring*

- Undertake to analyze IEC 60870-5, DNP3, and UCA 2.0 to see which one may suit their NS/EP and CIP missions best.
- Monitor and participate as appropriate in the IEEE standards process as it relates to SCADA systems, which are being developed through the IEEE Power Engineering Society.
- Participate in the ANSI-HSSP. This panel is looking into refining and creating standards critical to Homeland Security. They are looking at Utilities, in particular, which heavily utilizes SCADA systems.

  - Monitor and participate, as appropriate, in the IEC standards process as it relates to SCADA systems. More specifically, participate in the development of the UCA 2.0 specifications.
  - Additional conclusions and recommendations for SCADA systems.

# IEIA - H-SCADA Certification Series

**MODULE 2: Critical Infrastructure: Homeland Security and Emergency Preparedness (by Robert Radvanovsky)**

**Lesson 1:** *Chapter 1*

- Homeland Security Presidential Directives (HSPD)
- Combating Terrorism through Immigration Polices
- Homeland Security Advisory System
- National Strategy to Combat Weapons of Mass Destruction
- Management of Domestic Incidents
- Integration and Use of Screening Information
- National Preparedness
- Defense of the United States Agriculture and Food
- Bio-defense for the 21$^{st}$ Century
- Comprehensive Terrorist-Related screening Procedures
- Policy for a Common Identification Standards for Federal Employees and Contractors
- National Strategy for Maritime Security
- What is Critical Infrastructure?
- What is the Private Sector?
- What is Critical Infrastructure Protection?
- What is Critical Infrastructure Preparedness?
- Critical Infrastructure Functions and Infrastructure

**Lesson 2:** *Chapter 2*

- H.R. 4417
- Bill to Amend Immigration and National Act to Provide Permanent Authority "S" Visa (S.1424, Public Law No. 107-45)
- Enhanced Border Security and Visa Entry Reform Act of 2002 (H.R. 3525, Public Law No. 107-173)
- Immigration and Nationality Act of 1952 (U.S. Code Title 8)
- Real ID Act of 2005 (H.R. 418 and H.R. 1268, Public Law No. 109-13)
- Communications and Network Security (Communications Assistance for Law Enforcement Act (CLEA) (H.R. 4922, Public Law No. 103-414, 108 STAT, 4279)
- E-911 Implementation Act of 2003 (H.R. 2898), CISRA (S. 3067), (CISA) (S.1993)
- Cyberterrorism

- Infrastructure
- Domestic Safety and Security
- Economic and Financial Security
- Emergency Preparedness and Readiness
- Medical and Health Care Security
- Transportation Security (Includes Maritime Security)
- Hazardous Materials

**Lesson 3:** *Chapter 3, Chapter 4 and Chapter 5*

- National Response Plan (NRP)
- National Incident Management Systems (NIMS)
- Incident Command Systems (ICS)

**Lesson 4:** *Chapter 6, Chapter 7 and Chapter 8*

- Emergency Preparedness and Readiness (EMR)
- Awareness Level Guidelines First Responders, Hazmat, DOT
- Level A: Operations Level
- Level B: Technician Level
- Security Vulnerability Assessment (SVA)
- Standards and Guidelines: National Fire preventions Association, North American Electric Reliability Council, American Gas Association, Instrumentation, Systems and Automation Society, American Petroleum Institute, Chemical Industry Data Exchange, HIPPA, PSQIA, GLBA, Sarbanes-Oxley Act, ANSI, FIPS

# IEIA - H-SCADA Certification Series

### Lesson 5: *Chapter 9, Chapter 10 and Chapter 11*

- Information Sharing and Analysis Centers (ISAC)
- Supervisory Control and Data Acquisition (SCADA)
- War Driving and Walking
- Critical Infrastructure Information (CII)
- Need-to-Know
- "For Official Use Only" (FOUO)
- Unclassified Controlled Nuclear information (UCNI)
- Critical Energy Infrastructure Information (CEIT)

## MODULE 3: Advances in Computers: Volume 71 Nanotechnology (Edited by Marvin V. Zelkowitz)

### Lesson 1: *Programming Nanotechnology: Learning from Nature*

- Introduction and Development in Nanotechnology
- Benefits of Computer Science for Nanotechnology
- Swarm Intelligence
- Perceptive Particle Swarm Optimization
- Perceptive Particle Swarm Optimization for nanotechnology
- Self-assembling Nanotechnology

### Lesson 2: *Nanobiotechnology: An Engineer's Foray into Biology*

- Introduction and Nanofabrication
- Nano-biotechnologies for Sensing and Actuating
- Nano-biotechnology for Drug Delivery, Therapeutics and Other Uses

### Lesson 3: *Toward Nanometer-Scale Sensing Systems, Natural and Artificial Noses as Models for Ultra-Small, Ultra-Dense Sensing Systems*

- Introduction and the Physiology of the Sense of Smell
- Electronic Noses: Chemical sensing Systems
- Nanosensors
- Designing a Nanometer-Scale Nose-Like Sensing System

## Lesson 4: *Simulation of Nanoscale Electronic Systems*

- Introduction and Simulation Hierarchy for Semiconductor Devices
- Simulation Issues in Nanoscale Silicon Devices
- Organic Molecular Devices
- Simulation of Molecular Condition
- Carbon Nanotubes
- Ionic Channels and Conclusions

## Lesson 5: *Identifying Nanotechnology in Society*

- Introduction and Definitions *ad infinitum* (and More Politics)
- Perspectives and Conclusions from Science

## Lesson 6: *The Convergence of Nanotechnology, Policy and Ethics*

- Introduction and Converging Paths
- From Convergence to Collaboration and their Conclusions

**MODULE 4: Design for Manufacturability and Yield for Nano-Scale CMOS (by Charles C. Chiang and Jamil Kawa)**
Design for Manufacturability and Yield for Nano-Scale CMOS
Charles C. Chiang and Jamil Kawa, ISBN: 978-1-4020-5187-6
Springer, Netherlands © 2007

## Lesson 1: *Chapter 1, Chapter 2, Chapter 3 and Chapter 4*

- What is DMF/DFY
- Random Defects
- Mathematical Formulation of Approximation Method
- Improving Critical Area
- Systematic Yield–Lithography
- Systemic Yield-Chemical Mechanical Polishing (CMP)

**Lesson 2:** *Chapter 5, Chapter 6, Chapter 7 and Chapter 8*

- Variability and Parametric Yield
- Design for Yield Wafer Level Variability
- Environmental Variability and Aging
- Parametric Yield
- Static Timing and Power Analysis
- Critical Path Method (CPM)
- Yield Prediction, Modeling, Enhancement Mechanisms and IP Development
- Case fri DFM/DFY Driven Design
- DFM/DFY EDA Design Tools

**MODULE 5: FCC ONLINE Frequencies (47 C.F.R. Statue 2.106) by FCC Office of Engineering and Technology**

**Lesson 1:** *Familiarization with Frequency Allocations as published by the Federal Register*
- Radiofrequency
- International
- Radionavigation, Maritime Mobile, Fixed Mobile
- Standard Frequency and Time Signal
- Satellite, Aeronautical Radionavigation, Space
- Fixed, Standard Frequency and Broadcasting
- Aviation, Mobile and Fixed

**Lesson 2:** *Frequency Code Familiarization Exercises*
- Frequencies below 9 KHz
- Frequencies above 9 KHz
- Terahertz
- Various Categories of Service
- Additional Allocations

## MODULE 6: FDA/OSEL Annual Report 2007 (by Food and Drug Administration's Center for Devices and Radiological Health)

### Lesson 1: *Laboratory Descriptions*
- Active Materials
- Biological Risk Assessment
- Biomolecular Mechanisms
- Biotechnology
- Cardiovascular and Intervention Therapeutics
- Electrical Engineering
- Electromagnetic and Wireless Technologies
- Electrophysiology and Electrical Stimulation
- Fluid Dynamics
- Image Analysis
- Imaging Diagnosis
- Ionizing Radiation Measurements Laboratory Materials, Performance
- Optical Diagnostics, Therapeutic and Medical Nanophotonics
- Radiation Biology, Software, and Systems Engineering
- Toxicology, Ultrasonics and CDRH Standards Management Program

### Lesson 2: *OSEL 2007 Laboratory Accomplishments*
- Biological Risk Assessment
- Biomolecular Mechanisms
- Electromagnetic and Wireless Technologies
- Electromagnetic and Wireless technologies
- Electrophysiology and Electrical Stimulation
- Fluid Dynamics
- Image Analysis and Diagnostics
- Optical Diagnostics, Therapeutic and Medical Nanophotonics
- Software, Toxicology and Ultrasonics
- Interagency Agreements and CRADAs

**MODULE 7:** *Guidelines on the Evaluation of Vector Network Analyzers (VNA) EURAMET (by European Association of National Metrology Institutes)*

**Lesson 1:** *Guidelines on the Evaluation of Vector Network Analyzers (VNA)*

- Introduction, Documentation, Reference Standards, Beadless Airlines and Calibration Kits
- Traceability Kit
- Mathematical Models and Calibration
- Uncertainty Evaluation

**Lesson 2:** *Uncertainty Evaluation for On-Port Measurements, UVRC*

- Magnitude, Effective Directivity, Effective Test Port Match and System Repeatability
- Connector Repeatability, Cable Flexure and Ambient Conditions
- Uncertainty Evaluation for Two-Port Measurements, $U_{VRC}$, $U_{TM}$

**MODULE 8:** *Telemedicine and e-Health: Abstracts from the American Telemedicine Association (by Eighteenth Annual International Meeting and Exposition on Telemedicine Mary Ann Liebert, Inc. Publishers)*

**Lesson 1:** *Foundation for Starting a Telemedicine Program*

- International Concerns of Telemedicine
- Cost Reduction
- Business Enterprises
- Designing for Scale, Mobile Apps and Technologies
- Teledermatology practice Guidelines and Pearls
- Surgical Tele-Monitoring
- Multi-State Telehealth
- Remote Neurocognitive Assessment: Military and Civilian Projects

## Lesson 2: *Evaluation Methods, Concepts and Next Wave-Competing*

- University-Based Telehealth
- Telemedicine Value Proposition, ROI and Sustainability
- Creating Better Disease Management for Diverse Populations
- Neurology Telemedicine and Nimbility
- Telemedicine and Humanitarian Aid
- Legal and Regulatory Issues of Telemedermatology

## MODULE 9: Section 8: Health Effects Associated with Exposure of Industrial Workers to Radiofrequency Waves

### Lesson: RF Toolkit

- Familiarize with the Rules and Regulations of Exposure
- Who Enforces the Exposure Regulations
- Who is Most Susceptible to RF

## MODULE 10: Sandia Report: SAND2004-5625 Unlimited Release Printed November 2004 Microfabrication with Femtosecond Laser Processing (A) Laser Ablation and Ferrous Alloys, (B) Direct-Write Embedded Optical Waveguides and Integrative Optics in Bulk Glass.

### Lesson: *Fermtosecond Laser, Ferrous Alloys, Microfabrication of Optical Waveguides and Integrated Systems*

- Femtosecond Laser
- Varying Substrate Angle on Feature Quality
- Optical Waveguides and Integrated optics in Bulk Glass
- Accelermoter
- Micromachining of Energetic Material Films
- Y- Coupler
- Directional Coupler
- Mach-Zehnder Interferometer
- Foturan Glass
- Photoluminescence Excited by 800nm Femtosecond Laser Pulses

**IEIA - H-SCADA Certification Series**

**MODULE 11: Nano Air Vehicles A Technology Forecast**

**Lesson:** *Design, Function and Monitoring of Nano Air Vehicle*

- Defense Advanced Research Projects Agency
- Remote-Controlled Nano Air Vehicle (NAV)
- Military Intelligence Indoors and Outdoors on the Urban Battlefield
- NAV Delivery Payload(s)
- Role of Optimal Monitoring and Communication

**MODULE 12: Supplemental Articles**

**Lesson: Various Articles that Discuss Current Technology Interface of SCADA**

- Extremely Scaled Silicon Nano-CMOS Devices
- Designs for Ultra-Tiny, Special-Purpose Nanoelectronic Circuits
- General Receipt for Flatbands in Photonic Crystal Waveguides
- Implantable Telemetry Platform System for In Vivo Monitoring of Physiological Parameters
- Viral Structure and Mechanics
- Designs for Ultra-Tiny, Special-Purpose Nanoelectronic Circuits

# IEIA - H-SCADA Bio-Energy Field Professional Study Guide for Examination

## MODULE 1

### History of SCADA Systems and Common System Components

**TEXT:**

**NCS TIB 04-1: National Communications System**
Technical Information Bulletin 04-1
Supervisory Control and Data Acquisition (SCADA) Systems
Office of the Manager, National Communications System, P.O. Box 4052
Arlington, VA 22204-4052 © October 2004

Module 1 begins with NCS TIB 04-1: National Communications System (NCS), SCADA System regulations, which examines potential answers to the question: What is SCADA?

The course project, which is also assigned in this module, focuses on the NCS TIB 04-1. You may find it helpful to read this text in conjunction with the other texts and papers in this course to complete the project.

After successfully completing this module you will be able to:

- Identify and describe National Communications System (NCS) SCADA systems
- Discuss TIB Analysis, Observations and Conclusions and
- Briefly state NCS role in SCADA Systems, Utilities, IEEE and Monitoring.

This module includes lessons, project assignments and progress test(s).

# IEIA - H-SCADA Certification Series

## MODULE 1

### LESSON 1: History of SCADA Systems and Common System Components

The first lesson traces the history of SCADA Systems as an introduction to the concepts of, component make-up, evolution through monolithic, distributed and networked evolution.

The goal of this Technical Information Bulletin (TIB) is to examine Supervisory Control and Data Acquisition (SCADA) Systems and how they may be used by the National Communications System (NCS) in support of National Security and Emergency Preparedness (NS/EP) communications and Critical Infrastructure Protection (CIP). An overview of SCADA is provided, and security concerns are addressed and examined with respect to NS/EP and CIP implementation. The current and future status of National, International, and Industry standards relating to SCADA systems is examined. Observations on future trends will be presented,. Finally, recommendations on what the NCS should focus on with regards SCADA systems and their application in an NS/EP and CIP environment are presented.

After successfully completing this lesson you will be able to:

- Discuss the SCADA systems.
- Explain the network evolution.
- Examine the protocols and systems used.

### Assignment

For Lesson 1, read pages 1-19 NCS: Technical Information Bulletin 04-1 Supervisory Control and Data Acquisition (SCADA) Systems. Complete your reading assignment as outlined below. DO NOT SUBMIT ANSWERS TO IEIA for grading.

1. The National Communications System (NCS) was established through a Presidential Memorandum signed by _____ on August 21, 1963.

    A. President Ronald Reagan.    C. President George W. Bush.
    B. President Abraham Lincoln.  D. President John F. Kennedy

2. _____ is a service that must provide voice band service in support of presidential communications.

   A. Voice Band Service
   B. Scaleable Bandwidth
   C. Nationwide Coverage
   D. Survivability

3. SCADA is an acronym for _____.

   A. Supervisory Control and Data Acquisition.
   B. State Controlled and Disease Agency.
   C. Satellites Concepts Disorder and Augmentation.
   D. Space Center Data Agency.

4. SCADA systems consist of all of the following except:

   A. One or more field data interface devices, usually RTUs, or PLCs, which interface to field sensing devices and local control switchboxes and value actuators.
   B. A communications system used to transfer data between field data interface devices and control units and the computers in the SCADA central host. The system can be radio, telephone, cable, satellite, etc., or any combination of these.
   C. A central host computer server or servers (sometimes called a SCADA Center, master station, or Master Terminal Unit (MTU).
   D. A collection of standard and/or custom software (sometimes called Human Machine Interface (HMI) software or Man Machine Interface (MMI) software) systems used to provide the MOPETT central host and operator terminal application, support the communications system, and monitor and control remotely located field data interface devices.

5. Field data interface devices form the " _____ " of a SCADA system.

   A. "Nose and throat"
   B. "Arm and leg"
   C. "Eyes and ears"
   D. "Hands and feet"

6. Remote Telemetry Units (RTUs) are primarily used to convert electronic signals received from field interface devices into the language (known as the _____) used to transmit the data over a communication channel.

    A. Equipment standard.
    B. Programmed intelligence.
    C. Communication protocol.
    D. Field conditions.

7. A PLC is a devices used to _____ and control of industrial facilities.

    A. System control
    B. Automate monitoring
    C. Remote telemetry
    D. Field data

8. As PLCs were used more often to replace relay switching logic control systems, _____ was used more and more with PLCs at the remote sites.

    A. Pneumatic controllers.
    B. Switching logic.
    C. Telemetry.
    D. Local control problem.

9. The _____ is intended to provide the means by which data can be transferred between the central host computer servers and the field-based RTUs.

    A. Remote field system
    B. Computer servers
    C. Dial-up
    D. Communication network

10. The Communication Network medium used can either be _____, _____ or _____.

    A. Clock, remote controller or server.   C. Heater, refrigerator or watch
    B. Cable, telephone or radio.            D. Alarm, monitor or antenna.

11. Remote sites are usually not accessible by _____.

    A. Roads.
    B. Telephone lines.
    C. Railroad rails.
    D. Mobile phones.

12. Historically, SCADA networks have been dedicated networks; however, with the increased deployment of office _____ and _____ as a solution for interoffice computer networking, there exists the possibility to integrate SCADA _____ into everyday office computer networks.

    A. LANs and WANs, SCADA LANs
    B. WiFi and SWIFT, Nomads
    C. SCADA and Interpol, WASP
    D. CMOS and MITRI, Super Systems

13. The _____ or master station is most often a single computer or a network of computer servers that provide a man-machine operator interface to the SCADA system.

    A. LAN/WAN
    B. RTU
    C. Central host computer
    D. SCADA

14. Host Computer platforms characteristically employed _____ - based architecture, and the host computer network was physically removed from any office-computing domain.

    A. COTS
    B. CMOS
    C. LAN
    D. UNIX

15. Today many SCADA systems can reside on computer servers that are identical to those servers and computers used for traditional office applications. This has opened a range of possibilities for the linking of SCADA systems to office-based applications such as the following:

    A. GIS systems
    B. Hydraulic modeling software
    C. Drawing management systems, work scheduling systems, and information systems.
    D. All of the above

# IEIA - H-SCADA Certification Series

16. A wide range of _____ (COTS) software products are available to develop specific configurations of SCADA hardware platform and SCADA system.

    A. College Office Technologies Scales
    B. Commercial off-the-shelf
    C. CI ON Technologies Systems
    D. Cis Organic Transfer Supers

17. What is the definition for communications protocol drivers?

    A. Software that allows engineering staff to configure and maintain the application hosted with the RTUs or PLCs.
    B. Software used to control the central host computer hardware.
    C. Software that handles the transmittal and reception of data to and from the RTUs and the central host.
    D. Software that is usually based within the central host and the RTUs, and is required to control the translation and interpretation of the data between ends of the communications links in the system.

18. SCADA Architectures have evolved in parallel with the growth and sophistication of modern computing technology. Currently, has gone through three generations of development. What are they?

    A. Monolithic, Prehistoric and National
    B. Systems, MITRI and CMOS
    C. Monolithic, Distributed and Networked
    D. Pre-Quantum, Networked and Guided

19. The connective SCADA master station itself was very limited by the system vendor. Connections to the master typically were done at the bus level via a proprietary adapter or controller plugged into the _____ back plane.

    A. Central Processing Unit (CPU)
    B. Targeted Integrated Universal
    C. SCADA Planner Plugs
    D. Nano Payload Guidance Darts

# IEIA - H-SCADA Certification Series

20. The First Generation SCADA architectures were served as operator interfaces providing the _____ for system operators.

    A. RTUs
    B. LAN
    C. Multiple stations mini computer class
    D. Human-machine interface (HMI)

21. Second Generation SCADA architectures were distribution of system functionality across network-connected systems served to increase power. The _____ was used to communicate with devices in the field.

    A. HMI
    B. LAN
    C. RTU
    D. WAN

22. The current generation of SCADA master station architecture is closely related to that of the _____ generation, with the primary difference being that of a/an _____ rather than a vendor controlled, proprietary environment.

    A. First, RTU utilizing protocols
    B. Second, open system architecture
    C. Third, utilizing open standards
    D. LAN, HMI vendor-proprietary

23. The "open" or "off-the-shelf" systems utilized for SCADA vendors have gradually gotten out of the hardware development business. Some of these system vendors were _____, _____ and _____.

    A. Compaq, Hewlett-Packard and Sun Microsystems
    B. Microsoft, Apple and CISCO
    C. IBM, Samsung and Digital
    D. ECATTS, FUGI and Kodak

24. SCADA Protocols allow the RTU to accept commands to operate control points, set analog output levels and respond to requests. Each SCADA protocol consists of _____.

    A. IEC 6870-5-101
    B. Distributed Network Protocol versions (DNP3)
    C. Two message sets or pairs
    D. Different layers of neutral control measures.

25. SCADA enhanced performance architecture utilizes Application Layers (OSI Layer 7), link interface and physical interface. Each of these layers of architecture is based on RTUs, meters, relays, and other _____.

    A. Intelligent Electronic Devices (IEDs)
    B. Synchronizing Data Frames (SDFs)
    C. International Telecommunications Union (ITUs)
    D. Electronics Industries Association (EIA)

26. The _____ utilizes the following basic application functions, as defined in IEC 60870-5-5 (1995-06), within the user layer; station initialization; cyclic data transmission; general interrogation; command transmission; data acquisition by polling; acquisition of events; parameter loading; file transfer; clock synchronization; transmission of integrated totals and test procedures.

    A. IDS to the RTU
    B. ASDU field length
    C. 60870-5-102
    D. Standard 101 Profile

**Lesson 2: TIB Analysis, Observations and Conclusions**

SCADA systems began in 1960's and have evolved as technology changes. The SCADA systems' evolutionary role with mainframe-based to client/server architectures and common communications protocols like Ethernet, Internet, and TCP/IP to transmit data from the field to the master control unit will be studied. SCADA protocols evolved from closed proprietary systems, to open system allowing designers to choose equipment and transmission devices, such as telephone, mobile phone, microwave and satellites. This section of the SCADA Study Guide will allow for the following:

## IEIA - H-SCADA Certification Series

- Explain Structure of ADSUs in IEC 60870-5-101.
- Discuss common DNP3 architectures in use today.
- Discuss auxiliary hardware such as cables, phones and other systems.

**Assignment**

For Lesson 2, read pages 19- 40 NCS: Technical Information Bulletin 04-1 Supervisory Control and Data Acquisition (SCADA) Systems. Complete your reading assignment as outlined below. DO NOT SUBMIT THEM TO IEIA for grading. You may check your answers with the key provided at the end of the lesson.

1. DNP3 was developed with the following goals:

   A. High data integrity
   B. Flexible structure
   C. Multiple Applications
   D. A, B and C

2. The DNP3 can also be used with several physical layers, and as a layered protocol is suitable for operation over local and some wide area networks such as the following:

   A. Minimized overhead
   B. Open standard
   C. A and B
   D. Central operations

3. In DNP3 rules for substation computers and master station computers to communicate data and control commands, the substation computer gathers data for transmission to the master computer such as:

   A. Binary Input data
   B. Analog and Count Data
   C. Files that contain configuration data
   D. A, B, and C

4. In DNP3 terminology, the element numbers are called the _____.

   A. Point indexes          C. Master data
   B. Zero-based             D. None closed loop systems.

# IEIA - H-SCADA Certification Series

5. The typical system architectures, where DNP3 is used shows common system architectures that are used today. The physical connection between one master station and one slave has a physical connection between the two is typically a dedicated or _____ .

   A. Wi-Fi computer
   B. Cell phone
   C. Dial- up telephone line
   D. Satellite

6. The second type of system is known as a _____. One master station communicates with multiple slave devices. Conversations are typically between the client and one server at a time.

   A. Multidrop design
   B. Data concentrator
   C. Hierarchial
   D. Slaves non-hearing

7. In recent years, several vendors have used Transport Control Protocol/Internet Protocol (TCP/IP) to transport DNP3 messages in lieu of the media as data in its data base. The link layer frames utilized in these systems are embedded into _____ packets.

   A. TCP/IP packets
   B. CRC packets
   C. Micro array packets
   D. DNP3 CRC packets

8. DNP3 devices sent to the frame, and data link control information, which then allows every frame to begin with two sync bytes that help the receivers determine where the frame begins. The length specifies the number of octets in the remainder of the frame, not including _____ octets. The link control octet is used between sending and receiving link layers to coordinate their activities.

   A. DNPE and TCP
   B. TCP/IP
   C. LCO
   D. Cyclical Redundancy Check (CRC)

9. One often hears the term "link layer confirmation" when DNPs is discussed. A feature of DNP3's link layer is the ability of the _____ of the frame to request the receiver to confirm that the frame arrived.

   A. Transmitter
   B. Receiver
   C. Payload
   D. Actuator

10. In DNP3, the term _____ is used with data and refers to the current value.

   A. Events
   B. Break
   C. Static
   D. Ground

11. _____ events occurs when a binary input changes form an "on" to an "off" state or when an analog value changes by more than its configured deadband limit.

   A. LANS
   B. DNP3
   C. WANS
   D. CMOS

12. When a DNP3 server transmits a message containing response data, the message identifies the object number and variation of every value within the message. Object and variation numbers are also assigned for all of the following:

   A. Counters, binary inputs, controls and analog outputs
   B. Systems, networks, controls and receivers
   C. Implementation levels, users and groups
   D. Index number X, index number Y and frames

13. DNP3 organizes complexity into three levels. The lowest level, level 1, only very basic functions must be provided and all others are optional. Level 2 handles more functions, objects and variations, and level 3 is even more _____ .

   A. Precise
   B. Transparent
   C. Generic
   D. Sophisticated

14. There are many different ways SCADA systems may be deployed. One way is through the twisted-pair metallic cable utilized by Telephone Company and contain a number of pairs of conductor. One advantage to this deployed system is no licensing and fewer approvals. Several others are the following-except:

    A. Existing pole infrastructure
    B. Subject to water ingress
    C. Economical for short distances
    D. Relatively high channel capacity (up to 1.54 MHs) for short distances

15. Coaxial metallic cable is constructed of a center copper conductor, polyvinyl chloride (PVC) insulation, a braided or extruded copper shield surrounding the center conductor and PVC insulation and a plastic jacket cover. Coaxial cable can transmit _____ signals up to several MHz with low attenuation compared to twisted pair wires used for telephone service.

    A. Low frequency
    B. Terahertz frequency
    C. High frequency
    D. Microwave frequency

16. Coaxial Cable disadvantages are all except the following:

    A. More immune to Radio Frequency (RF)
    B. Right-of-way clearance required for buried cable
    C. Subject to ground potential rise due to power faults and lightening
    D. Inflexible network Configuration

17. Optical fibers consist of an inner core and cladding of silica glass and a plastic jacket that physically protects the fiber. Two types of fibers are usually considered: _____ and _____ fiber .

    A. Polyethylene and single-mode
    B. Nylon multi-mode and polyvinylchloride
    C. A and B
    D. Multi-mode graded index and single-mode step index

18. The use of fiber optics in SCADA systems have all of the following advantages except:

    A. Located where the circuits are required
    B. Digital PLC has capacity for three to four channels (two voice and one high speed data.
    C. Analog PLC has capacity for two channels (one voice and on" speech plus" low speed data)
    D. Inherently few channels available

19. The use of satellites are positioned in geo-stationary orbits above the earth's equator and thus offer continuous coverage over a particular area of the earth with its "foot print" for the satellite. Earth stations are comprised of an _____ pointing at the satellite, a radio transceiver with a low-noise amplifier, and baseband equipment.

    A. Radio
    B. Eyeglass transmitter
    C. Ku-band receiver
    D. Antenna

20. Satellites use both the C-band and the Ku-band, Very Small Aperture Terminal (VSAT) technology has advanced to the point where a much smaller antenna (down to about _____ meter(s)) can be used for Ku-band communications.

    A. 1
    B. 100
    C. 1000
    D. 10,000

21. Leased telephone circuits have long been used to meet communications needs. Most organizations use standard telephones connected to the _____ for office communications and for routine voice traffic to stations. Leased dedicated circuits are used for dedicated communication requirements, such as telemetry and SCADA.

    A. International System Network (ISN)
    B. Public Switched Network (PSN)
    C. Satellite Guided Network (SGN)
    D. Telecommunications Guidance Network (TGN)

22. Leased telephone lines utilize wideband channels for high speed data signaling. Circuit characteristics can often be conditioned for many other uses including _____ and various types of low and medium speed data.

    A. Voice
    B. Hearing
    C. Smelling
    D. Touch

23. Very High Frequency (VHF) band extends from 30 to 300 MHz and is usually used by utilities for mobile radio, although point-to-point links have been implemented in this band for joint voice and data use, such as used in _____ and _____ dispatching systems.

    A. Teacher and student
    B. Ice-cream truck and child
    C. Taxi and police
    D. Building to mobile unit

24. Ultra High Frequency Radio (UHF) band extends from 300 to 3000 MHz. The bands typically considered or UHF radio are in the 400 MHz and 900 MHz range. Most of the radio products for SCADA applications are available in the U.S. Operate in the 900 MHz frequency range. The _____ regulates the use of radio frequencies and has designated the 928 and 952 MHz range specifically for use by utilities for data communication applications.

    A. Multiple Address Radio Systems (MARS)
    B. Federal Communications Commission (FCC)
    C. Point-to-Multipoint (PTM)
    D. U.S. Congress

25. Point-to-point communications is usually used for SCADA communications from a master station or dispatch center to individual substations. Their disadvantages are all of the following except:

    A. Less stringent waveguide and antenna requirements
    B. Low channel capacity
    C. Low digital data bit rate
    D. Limited transmission techniques available

26. A Multiple Address Radio System (MARS) Radio system generally consists of one Master Station (usually Hot standby, full duplex) transmitting over an omni directional, gain antenna in a 360 radiation pattern to fixed station remotes or slaves (usually Non-standby, half duplex) that receive the signals via a directional, gain antenna. The _____ MHz Mars radio is a single channel system that communicates with each of its remotes or slaves in sequence. Services usually supported by MARS are SCADA, Telemetry/Data Reporting, and voice (on a limited basis).

A. 78/900 MHz
B. 12.5/83 MHz
C. 400/900 MHz
D. 5.9/1000 MHz

27. Low power spread spectrum radios are allowed to operate in the 9802-928 MHz band, 2.4 and 5.3 GHz band without licenses. This has promoted the development of packet-type radio networks for data systems, which are appropriate for _____ applications, such as Distribution Automation (e.g. utilities).

A. Spread Spectrum Radio System (SSRS)
B. Digital Multiples System (DMS)
C. Federal Communication Commission (FCC)
D. Sun Micro Systems (SMS)

28. Microwave radio is a term used to describe UHF radio systems operating at frequencies above 1 GHz., although multi-channel radio systems operating below 1 GHz are sometimes referred to as microwave systems. These systems have high channel capacities and data rates, and they are available in either _____ or _____ transmission technologies.

A. Micro, terahertz
B. Analog, digital
C. Point, multiple point
D. UHF, VHF

29. Point-to-multiple point radio systems can operate in several modes except:

A. Frequency Division Multiple access (FDMA)
B. Time Division Multiple Access (TDMA)
C. Code Division multiple Access (CDMA)
D. Microwave Radio Access (MRA)

**Lesson 3: NCS role in SCADA systems, utilities, IEEE and Monitoring**

SCADA systems are very diverse and precise. One will undertake to analyze IEC 60870-5, DNP3 and UCA 2.0 to see which one may suite their NS/EP and CIP missions' best applications. Some of the topics may overlap into a larger expansion of topics discussed in Lesson 2. The participant will address ANSI-HSSP, IEEE Power Engineering Society and their roles in SCADA and in creating standards critical to Homeland security.

**Assignment**

For Lesson 3, read pages 41- 64 NCS: Technical Information Bulletin 04-1 Supervisory Control and Data Acquisition (SCADA) Systems. Complete your reading assignment as outlined below. DO NOT SUBMIT THEM TO IEIA for grading.

1. SCADA systems have evolved in recent years and are now based on open standards and _____ products.

    A. COTS
    B. TCP
    C. IPP
    D. PLC

2. On October 1, 2003, _____, Director, Information Security Issues at the general Accounting Office (GAO) eluded to the security and vulnerability of SCADA systems and other issues in his testimony before the subcommittee on Technology, Information Policy, Intergovernmental Relations and the Census, House Committee on Government Reform.

    A. George Burns
    B. Mickey Mantel
    C. Robert Dacey
    D. George Bush

3. SCADA Administrators and _____ are often deceived into thinking that since their industrial networks are on separate systems from the corporate network they are safe from outside attacks.

    A. Industrial Systems Analyst    C. Environmental Engineers
    B. Epidemiologist                D. Bio-Systems Engineers

## IEIA - H-SCADA Certification Series

4. Security in an industrial network can be compromised in many places along the system and is most easily compromised at the SCADA host or control room level. Some of the types of attacks the SCADA host may experience are all except the following:

   A. Use a Denial of Service (DOS) attach to crash the SCADA server
   B. Delete system files on the SCADA server
   C. Plant a Trojan and take complete control of the system
   D. Use SCADA server as a non IP Spoofing network

5. To develop an appropriate SCADA security strategy involves analysis of multiple layers of both the corporate network and SCADA architectures which include all of the following:

   A. Firewalls, proxy servers, operating systems    D. A, B and C
   B. Application system layers
   C. Communications, policy and procedures

6. A _____ is when firewalls, properly configured and coordinated, can protect passwords, IP addresses, files and more.

   A. Operating systems         C. Border Router and Firewalls
   B. SCADA Firewalls           D. Proxy Corporate Router

7. _____ are used for layer attacks; i.e., buffer overruns, worms, Trojan Horse programs and malicious Active-$X^5$ code, can incapacitate anti-virus software and bypass the firewall as if it wasn't even there.

   A. SCADA firewalls           C. Operating Systems
   B. Applications              D. Proxy Servers

8. Facilities using Wireless Ethernet and Wired Equivalent Protocol (WEP) should change the default name of the Service Set Identifier (SSID) when _____ should be segmented off into their own IP segment using smart switches and proper sub-masking techniques to protect the Industrial Automation environment from the other network traffic.

   A. SCADA Firewalls                  C. SCADA Base Systems
   B. SCADA Internal Network Design    D. SCADA Policies

## IEIA - H-SCADA Certification Series

9. In summarizing SCADA systems for security, the multiple "rings of defense" must be configured in a complementary and organized manner, and the planning process should involve a cross-team with senior staff support from which of the following:

    A. Hospital, fire department and homeland security
    B. President, military command and police
    C. Trojan horses, hard drive and worm holes
    D. Operations, facility engineering and Information Technology (IT)

10. The Institute of Electrical and Electronics Engineers (IEEE) is a membership organization that produces Electrical and IT-Related standards that are used internationally. Which of the following standards have been published by the IEEE with respect to SCADA Systems:

    A. IEEE Std 9999-192 – IEEE Recommended Practice for Master/Remote Supervisory Control and Data Acquisition (SCADA) communications
    B. IEEE Std 1379-2000 – IEEE Recommended Practice for Data Communications Between Remote Terminal Units and Intelligent Electronic Devices in Sub-station.
    C. IEEE Std ANSI 501( c) 3           D. A and B

11. The mission of the American National Standards Institute's Homeland security Standards Panel (ANSI-HSSP) is to identify existing consensus standards, or, if none exists, assist the _____ and those sectors requesting assistance to accelerate development and adoption of consensus standards critical to homeland security.

    A. Department of Environmental Protection
    B. Department of Occupational Safety and Health
    C. Department of Interior           D. Department of Homeland Security

12. The electric Power Research Institute (EPRI) was founded in 1973 as a non-profit energy research consortium for the benefit of utility members, their customers, and society. The EPRI has developed the _____ to integrate communications for "real-time" utility operations for SCADA systems.

    A. International Electrotechnical Commission (IEC)
    B. Utility Communications Architecture (UCA)
    C. TR1550 Volume 1 and 2
    D. Electric Power Trade Commission (EPTC)

## IEIA - H-SCADA Certification Series

13. IEC 60870-5 specifies a number of frame formats and services that may be provided by different layers used in SCADA systems.
    Who charted the development of this system?

    A. International Electrotechnical Commission (IEC)
    B. AOC-IOU         C. EPRI         D. Homeland Security

14. _____ is based on the standards of the International Electrotechnical Commission (IEC) Technical committee 57, Working Group 03, who have been working on an OSI 3 layers "Enhanced Performance Architecture " (EPA) protocol standard for telecontrol applications.

    A. RTU         D. EPRI
    B. DNP3        C. IED

15. Today's SCADA systems are able to take advantage of the evolution form mainframe based to _____ architectures. These systems use common communications protocols like Ethernet and TCP/IP to transmit data from the field to the central master control unit.

    A. Telephone lines/electrical    C. Client/server
    B. Director/ owner               D. Hijacking/system

16. SCADA Systems, like other _____ systems, are subject to many common security attacks such as viruses, denial of service and hijacking of the system.

    A. Computer    C. Guided
    B. Technical   D. Learning

17. While SCADA protocols are more open today, there is no clear consensus of which protocol is best. _____ series and DNP3 have many similarities but are not 100% compatible.

    A. UCA 2.0                  C. IEC 60870-5
    B. IEEE Technical Report    D. A and C

18. The _____ should monitor the development, and make contributes when appropriate, of IEC 60870-5, DNP3 and UCA 2.0.

    A. SCADA    C. NS/EP
    B. NCS      D. CIP

## IEIA - H-SCADA Certification Series

19. NCS should look at commissioning additional studies that examine unconventional attacks, such as those using _____ , against SCADA systems supporting NS/EP and CIP.

    A. Quantum Dot Recovery Weapons
    B. Non-lethal Weapons
    C. Ion Impulse System Weapons
    D. Electro Magnetic Pulse (EMP) Weapons

The following questions will consist of matching the appropriate acronyms with the appropriate meanings.

20. ADSS     _____

21. B-ISDN   _____

22. EPRI     _____

23. FCC      _____

24. HF       _____

25. HTTP     _____

26. DNP3     _____

27. CRC      _____

28. OMNCS    _____

29. RF       _____

30. RTU      _____

31. TCP/IP   _____

32. VHF      _____

33. WAN      _____

# IEIA - H-SCADA Certification Series

## IEIA - H-SCADA Bio-Energy Field Professional Study Guide for Examination

### MODULE 2
### Critical Infrastructure: Homeland Security and Emergency Preparedness
### (by Robert Radvanovsky)

Module 2 begins with the textbook: **Critical Infrastructure: Homeland Security and Emergency Preparedness** by Robert Radvanovsky, CRC Press/Taylor & Francis. Boca Raton, FL © 2006   ISBN: 0-84593-7398-0

The course project, which is also assigned in this module, focuses on the role of Homeland Security and Emergency Preparedness as it relates to SCADA systems. You may find it helpful to read this text in conjunction with the other tests and papers in this course to complete the project.

Some materials used within the referenced text book publication were taken in part or in their entirety from several very reliable and useful sources. Any information that may appear to be repetitive in its content from those sources was taken to provide a more introspective perception of what defines, "critical infrastructure preparedness."

The United States Department of Homeland Security:  www.dhs.gov

The Office of Domestic Preparedness (ODP) is part of the United States Department of Homeland Security's Office of State and Local Government Coordination and Preparedness (SLGCP):  www.ojp.usdoj.gov/odp

The Federal Emergency Management Agency (FEMA) is part of the United States Department of Homeland Security's Emergency Preparedness and Response Directorate:  www.fema.gov

The United States Fire Administration is part of FEMA, which is part of the United States Department of Homeland Security's Emergency Preparedness and Response Directorate:  www.usfa.fema.gov

The National Fire Prevention Association:  www.nfpa.org

After successfully completing this module you will be able to:

- Know the procedures and guidelines provided by President Clinton's President Decision Directive called "Critical Infrastructure," PDD-63.

- Understand key factors, terms and guidelines granted to Homeland Security and Emergency Preparedness foundations.

- Understand and apply the fundamentals behind securing, protecting, and safekeeping our nation's infrastructures – all relevant industries, national landmarks, and national assets – that are considered critically vital to the continued economic success and operation of the United States.

# IEIA - H-SCADA Certification Series

## MODULE 2

**Lesson 1**

**Chapter 1: Introduction to Critical Infrastructure Preparedness**

The first lesson traces the history of SCADA Systems through the creation of Homeland Security, Emergency Preparedness and Global Domestic Security. It is an introduction to the concepts for the base of the entire text book and what is described in some of the historical backgrounds of "critical infrastructure," and why it is important to the United States. There will be some terms and definitions covering a brief synopsis of the intent of the textbook, and what is to be expected from Critical Infrastructure Protection and Preparedness Specialist (CIPS) professional as it relates to global integration will be addressed. Read pages 1-14.

After successfully completing this lesson you will be able to:

- Identify and understand the role of Homeland Security Presidential Directives (HSPD)
- Define a Critical Infrastructure
- Distinguish Private and Public Sectors
- Define what is Critical Infrastructure Protection, Preparedness and/or Function

**Assignment**

For Lesson 1, read pages 1-10 textbook by Robert Radvanovsky, <u>Critical Infrastructure: Homeland security and Emergency Preparedness.</u> Complete your reading assignment as outlined below. DO NOT SUBMIT ANSWERS TO IEIA for grading.

1. Officials from both public and private sectors of the United States have been actively promoting and applying critical infrastructure protection and preparedness methods well before the attacks on _____. Yet it was not too long ago that most citizens never heard of such words, at least until _____.

    A. December 25, 2013; tiger teams
    B. September 11, 2001; 9/11
    C. July 4, 1776; tea party
    D. November 11, 2011; 11/11/13

# IEIA - H-SCADA Certification Series

2. In the past the concept of government or industry-sponsored teams of experts and professionals who attempt to break down any defenses or perimeters in an effort to uncover, and eventually patch, any securities holes were given the following terms except which one:

   A. Tiger teams
   B. Bad guys
   C. Red teams
   D. Rainbow crystals

3. Homeland Security Presidential Directives (HSPD) are presidential directives specifically relating to Homeland Security. Each directive has specific meaning and purpose and is carried out (almost immediately) by the _____.

   A. U.S. Department of Interior
   B. U.S. Federal Bureau of Investigation
   C. U.S. Department of State
   D. U.S. Department of Homeland Security

4. Homeland Security Council (HSC) was established on October 29, 2001. The HSC's main function is to do the following:

   A. Implement a corresponding "Protective Measures"
   B. Perform essentially security services
   C. Ensure coordination of all homeland security related activities among executive departments and agencies and to promote the effective development and implementation of all homeland security policies.
   D. Enhance the ability of the United States to manage domestic incidents by establishing a single, comprehensive National Incident Management System (NIMS).

5. _____ was established on October 29, 2001. This directive defined border and immigration security policies.

   A. HSPD-6
   B. HSPD-2
   C. HSPD-4
   D. HSPD-5

6. _____ was established on March 11, 2002 to address "Threat Conditions" and establish "Protective Measures."

   A. HSPD-3
   B. HSPD-4
   C. HSPD-6
   D. HSPD-5

# IEIA - H-SCADA Certification Series

7. _____ directive establishes a national policy to defend the agriculture and food system against terrorist attacks, major disasters, and other emergencies.

   A. HSPD-10
   B. HSPD-11
   C. HSPD-9
   D. HSPD-8

8. _____ this directive initiated methods of enhancing security, increasing government efficiency, reducing identity fraud, and protecting personal privacy by establishing a mandatory, government wide standard for security and reliable forms of identification issued by the federal government to its employees and contractors.

   A. HSPD-12
   B. HSPD-10
   C. HSPD-1
   D. HSPD-13

9. What is critical infrastructure?

   A. Refers to assets of physical and computer-based systems that are essential to the minimum operations of the economy and government.
   B. Includes, but are not limited to, telecommunications, energy, banking and finance, transportation, water systems and emergency services, both government and private.
   C. All areas and things of, on, under, relating to, adjacent to, or bordering on a sea, ocean, or other navigable waterway.
   D. A and B

10. _____ sector of a nation's economy consists of those entities which are not controlled by the state; that is, a variety of entities such as firms, companies, corporations, private banks, non-governmental organizations, etc.

    A. Public
    B. Universal
    C. Private
    D. Cosmic

11. The public sector consists of the following:

    A. Public health, health care, chemicals
    B. Federal, state, local and tribal governments
    C. Surrounding organizations, environmental and newspaper agencies
    D. CIPR, man-made materials and government

# IEIA - H-SCADA Certification Series

12. Critical Infrastructure Protection (CIP) is not only critical in protecting people and cyber systems for the private and public sector, but also aids in attacks involving all of the following except:

    A. Hazardous materials (HAZMAT) and nature
    B. Chemical substances, radiological and biological
    C. Music, voice and space
    D. Terrorist, other criminals, hackers and so forth

13. _____ includes reading and interpreting visually represented materials for risk mitigation (and its remediation, vulnerability assessments (as part of periodic assurance tests) and competency or accreditation courses for those who validate, protect and safeguard our infrastructures.

    A. Critical Infrastructure Preparedness
    B. Critical Infrastructure Functions
    C. Critical Infrastructure Protection
    D. Critical Infrastructure Public and Private Sector

14. The origins of Critical Infrastructure took place in the mid-1990's with specific infrastructure sectors (and its assets) falling under its definition of the Homeland Security Act of 2002 (P.L. 107-296, Sec. 2.4) are all the following Except which one:

    A. Information technology, telecommunications and chemicals
    B. Transportation systems, emergency services, postal and shipping services
    C. Agriculture and food, drinking water and water treatment
    D. Pet care, cultural foods and music

*Additional Study guide questions and notes by the author of this text, Robert Radvanovsky are available on text pages 8 - 11 of Chapter 1.*

## IEIA - H-SCADA Certification Series

**LESSON 2: Chapter 2: Regulations and Legislation**

The second lesson of Module 2 focuses on the laws, bills and regulations that were implemented after September 11, 2001 as they apply to SCADA systems.

After successfully completing this lesson you will be able to:

- Identify the categories of the laws.

- Understand border security and immigration laws.

- Identify Communications and Network Security.

- Understand the applications of Domestic, Aviation, Maritime and Public Security.

**Assignment**

For Lesson 2, read pages 13 – 40 of your textbook. Complete your reading assignments as outlined below. DO NOT SUBMIT ANSWERS TO IEIA for grading.

1. The following categories of the laws as defined by National Strategy identified as "critical infrastructure sector" are below except:

    A. Alien Transit Authority
    B. Hazardous Materials (includes Weapons of mass Destruction (WMD)
    C. Domestic Safety and security
    D. Cyber terrorism

2. The _____ has asked state and local police to voluntarily arrest aliens who have violated criminal provisions of the U.S. immigration and Nationality Act or civil provisions that render an alien deportable, and who are wanted as recorded by the National Crime Information Center.

    A. Enhanced Border Security and Visa Entry Act 1999
    B. Immigration and Nationality Act of 1952
    C. Communications Assistance for Law Enforcement Act
    D. E-911 Implementation Act of 2003

3. In the Real ID Act of 2005 (H.R. 418 and H.R. 1268) Public Law No. 109-13 the establishment of national standards for state issued drivers' licenses and non-drivers' identification cards were established. Asylum under this act may be granted/authorized by the Secretary of Homeland Security and the _____.

   A. The President of the U.S.
   B. The Inspector Secret Services
   C. The Attorney General
   D. The Local Law Enforcement Agency, Sheriff

4. The Communications Assistance for Law Enforcement Act (CALEA) was signed into law on October 25, 1994. This act amended the federal criminal code to make it clear that the following is true:

   A. Laws, bills, and regulations for protecting and safeguarding any devices are protected that are connected to our phone systems.
   B. Emphasizes the registration requirements from visiting foreign nationals from the 18 countries affected.
   C. A telecommunications carrier's duty to cooperate in the interception of communications for law enforcement purposes.
   D. Calls for information sharing among border security, law enforcement, and intelligence agencies among other provisions.

5. The Government Information Security Act of 2000 provided a comprehensive framework for establishing and ensuring the effectiveness of controls over information and information resources that support federal operations and assets. The _____ noted that the "cyber attacks incident to conflicts to the Middle East "emphasized the potentially disastrous effects that such concentrated attacks can have on information and other critical government and private sector electronic systems.

   A. E-Government Act of 2002
   B. Computer Security Enhancement Act of 1997
   C. Warren and Bush Investigation
   D. Gilmore Commission

6. The Computer Security Enhancement Act of 1997 (H.R. 1903) specifically requests the _____ to fulfill its responsibilities under the computer standards program to request from the private sector, assist in establishing voluntary interoperable standards and guidelines among other requirements.

    A. National Academy of Science
    B. National Intelligence
    C. National Institute of Standards and Technology
    D. National Securities and Trade Association

7. The Federal Information Security Management Act of 2002 (FISMA), authorizes and strengthens the information security program, evaluation, and reporting requirements for federal agencies. Its security objectives are the following:

    A. Confidentiality, integrity and availability
    B. Peace, law and order
    C. Imaging, profiling and observation
    D. Evaluating, describing and cultivation

8. Infrastructure pertains (mostly) to buildings, bridges and spans – essentially, the fabrication of an infrastructure. All of the Acts listed below pertain to infrastructure except:

    A. E-Government Act of 2002
    B. Energy Policy Act of 2005
    C. National Construction Safety Team Act
    D. Pipeline Safety Improvement Act of 2002

9. The following acts are identified as part of domestic safety and security under Homeland Security Act of 2002. One of the acts listed below is not included as Part of domestic safety and security. Identify the one that is not.

    A. Intelligence Reform and Terrorism Prevention Act of 2004
    B. National Intelligence Reform Act of 2004
    C. Preparedness Against Terrorism Act of 2000 (H.R. 4210)
    D. United States Wild Cat, Fire and Jet Act of 2013

10. The Intelligence Reform and Terrorism Prevention Act of 2004 creates an Intelligence Chief with broad authority to unify intelligence gathering and operations. Additional provisions are all of the following, except which item.

A. Authorize the FBI to conduct surveillance of foreign nationals suggested of Terrorism.
B. Will increase the number of detention spaces available for terrorists.
C. Do not create a uniform security-clearance process.
D. Allow federal prosecutors to share information from Grand Jury proceedings with law enforcement to prevent terrorist attacks.

11. The USA Patriot Act of 2001 (H.R. 3162, Public Law. No. 107-56) is organized by special topics, select the best response.

    A. Major speeches.
    B. Dispelling the myths about the act.
    C. Congressional votes and explanations of the act.
    D. All of the above

12. The Wireless Communications and Public Safety Act of 1999 (H.R. 438, S. 800) is a nationwide emergency number, limits commercial disclosure as well as the reuse of location information from mobile telephones. This number is _____.

    A. 911        B. 611        C. 411        D. "0" (operator – zero)

13. Which Act offers "Rewards" for Justice Program under which the U.S. Secretary of State may offer rewards for information that prevents or favorably resolves acts of international terrorism against U.S. persons or property worldwide?

    A. 2001 Emergency Supplemental Appropriations Act
    B. 1984 Act to Combat International Terrorism
    C. Wireless Communications and Public Safety Act
    D. Critical Infrastructure Resource Act of 2001

14. All of the following acts apply to the Emergency Preparedness and Readiness Act. Select the item that does not apply to this act.

    A. Community Protection Act of 2003
    B. Emergency Preparedness and Response Act of 2003
    C. First Responders Partnership Grant Act of 2003
    D. Terrorism Risk Insurance Programs Act of 2002

# IEIA - H-SCADA Certification Series

15. Project _____ - authorizes appropriations to develop procure and make available countermeasures against chemical, biological, radiological or nuclear weapons that could cause public health emergency affecting national security.

    A. D-TeK     B. Microbic ID     C. Bio Shield     D. UMRA

16. _____ Emergency Personal Protection Act of 2003, which authorizes benefits and other compensation for certain individuals with injuries resulting from administration of _____ countermeasures.

    A. Food-borne illness; phage
    B. Small pox; small pox
    C. DNA thiophene chip; thiols
    D. Polio; magnesium channel blockers

17. Transportation Security (includes) _____ Security and Airport Security.

    A. Aviation     B. Victim     C. Maritime     D. Port

18. HAZMAT handling is covered by the U.S. Department of Transportation, while Weapons of Mass Destruction (WMDs) investigations, there is the _____ of the United States regarding weapons of mass destruction.

    A. Maritime Transport Security              C. OSHA
    B. Commission on the Intelligence Capabilities     D. EPA

19. Chemical agents used as a weapon may include _____ assistance to enforce prohibition in certain emergencies.

    A. Military     B. Law Enforcement     C. Fireman     D. Public Safety

20. _____ legislation was passed to reduce the threat from new chemicals that present or will present an unreasonable risk of injury to health or the environment.

    A. Occupational Safety and Health Act
    B. Superfund Amendment and Reauthorization Act (SARA)
    C. Toxic Substance Control Act (TSCA) of 1986
    D. Chemical Security Act of 2005

21. This act was specifically developed to address "high priority category chemical source." The name for this act is called _____.

   A. Chemical Facility Act of 2001
   B. Superfund Amendments and Reauthorization Act (SARA)
   C. Toxic Substance Control Act (TSCA) of 1986
   D. Chemical Security Act of 2005

22. The Chemical Facility Security Act of 2005 requires the Secretary of _____ to designate certain combinations of chemical source and substance of concern as high-priority categories.

   A. Transportation
   B. Defense
   C. Homeland Security
   D. State

*Additional Study guide questions and notes by the author of this text, Robert Radvanovsky are available on text pages 34-40 of Chapter 2.*

**LESSON 3: Chapters 3, 4, and 5.  Read text pages 41 – 118.**

The third lesson addresses the National Response Plan (NRP), National Incident Management Systems (NIMS) and Incident Command Systems (ICS) aspects as issued by the Department of Homeland Security (DHS). It outlines at strategic levels the what, where and how in case of a crisis or large scale disaster, usually at a national level. Additional discussion and study focus notes are identified at the end of each chapter for additional technical support by the author of the text, Robert Radanovksy.  See the following sections of the text ~ Chapter 3: text pages 55-60 ~ Chapter 4: text 83 – 93 ~ Chapter 5: 114 – 118.

**Assignment**

For Lesson 3, read pages 41 - 118 of your textbook. Complete your reading assignments as outlined below. DO NOT SUBMIT ANSWERS TO IEIA for grading.

# Chapter 3

1. What is the National Response Plan (NRP)?

    A. A comprehensive all-hazardous approach to enhance the ability of the US to manage a nationwide template for working together to prevent or respond to threats and incidents regardless of cause, size or complexity.
    B. Protect the health and safety of the public.
    C. Conduct law enforcement investigations.
    D. Protect property and mitigate damage.

2. Who regulates NRP Training?

    A. OSHA
    B. EPA
    C. FEMA
    D. CONPLAN

3. NRP is broken down into all of the following subcategories except which one.

    A. Base Plan
    B. Appendixes
    C. Emergency Support Function Annexes and Fun
    D. Support and Incident Annexes.

4. NRP uses all of the following mechanism between NIMS except which one.

    A. Facilitate Federal to Field Incident and Emergency Support
    B. Critical Infrastructure/Key Resource
    C. Facilitate a monkey fund
    D. Improve coordination of government, private-sector and nongovernmental organizational partners.

5. _____ serves to primary national level multiagency hub for domestic situational awareness and operational coordination.

    A. National Response Coordination Center (NRCC)
    B. Joint Field Office (JFO)
    C. Principal Federal Official (PFO)
    D. Homeland Security Operations Center (HSOC)

6. The NPP uses the foundation provided by the Homeland Security Act, HSPD-5 and the _____ Act.

   A. Stafford
   B. Bush
   C. Kissinger
   D. Obama

7. The _____ is the Commander in Chief of State Military force (National Guard when in state active duty or Title 32 states and the authorized state militias).

   A. President
   B. Vice President
   C. Mayor
   D. Governor

8. The Secretary of _____ is the "principal official" for domestic incident management.

   A. State
   B. Homeland Security
   C. Defense
   D. Transporation

9. Military forces, command runs from the President to the _____ to the Commander of the Combatant Command to the Commander of the _____.

   A. Secretary of Defense; force
   B. Secretary of Transportation; strength
   C. Secretary of Energy; change
   D. Secretary of State; funds

10. The scope of Emergency Support Functions (EFSs) include all of the following except which one.

    A. Transportation
    B. Communications
    C. Fire Service
    D. International Affairs

## Chapter 4

1. _____ allows a template to enable federal, state, local and tribal governments and private sector and nongovernmental organizations to work together effectively to prevent acts of catastrophic terrorism.

    A. NPP
    B. DHS
    C. NIMS
    D. OSHA

2. The components of NIMS include all of the following concepts except which one.

    A. Command and Management
    B. Preparedness and Resource Management
    C. Commanders and Information Management
    D. Support technologies, ongoing management, maintenance and plant shut-downs.

3. NIMS is used to support technologies which include all except _____ functions.

    A. Voice and data communications systems
    B. Record keeping and resource tracking.
    C. Data display systems
    D. Non-specialized technologies

4. Incident Command Systems (ICS) is a management system for all of the following areas except which one.

    A. Command
    B. Operations
    C. Planning
    D. Killing

5. ICS requires the use of common terminology and use of "Clear Text," - that is, communications without the use of any organized specific codes, _____, or _____ and _____.

    A. Jargon
    B. Ciphers
    C. Plain, old, simple English
    D. All of the above

6. "Accountability" under ICR is the primary responsibility of control is under _____.

   A. Chain of command
   B. Provide check in capabilities for all
   C. Unity of Command
   D. Only one supervisor

**Match the proper response to the appropriate Unified Comment (UC) or ICS. Only 1 match for each statement. Questions 7 – 10.**

   A. UC (Unified Command)
   B. ICS (Incident Communication System)
   C. AC (Area Command)
   D. MACS (Multiagency Coordinate System)

7. Incidence Cross Political Jurisdiction _____

8. Geographically dispersed evolving over a period of time. _____

9. Direct tactical and operational responsibility. _____

10. Multiple concurrent incidents. _____

**Match the correct acronym with the proper name. Questions 11 – 14.**

   A. Emergency Operations Center          _____ 11. EOC
   B. Public Information Office            _____ 12. JIC
   C. Joint Information Systems            _____ 13. PIO
   D. Joint Information Center             _____ 14. JIS

15. JIC Organizational Structure for Level I; Level II and Level III. Are all correct statements except for which one.

   A. Level I: Joint Information Center (JIC)
   B. Level II: Press Security (Jurisdiction) and JIC Liaison (as needed)
   C. Level III: Research Team, Media Team Logistic Team
   D. Level III: Press, Food and Communications

16. Under Emergency Operations plan the personnel qualifications and certification for NIMS must include the following:

    A. Training, expenses
    B. Credentialing, currency
    C. A, B and D
    D. Physical and medical fitness

17. _____ and _____ provide the means for one jurisdiction to provide resources or other support to another jurisdiction during the incident.

    A. Multi-aid agreement
    B. EMACS
    C. SBDI
    D. A and B

18. _____ involves categorizing resource by capability based on measurable standards of capability and performance.

    A. Identity          C. Order
    B. Resource typing   D. Training

## Chapter 5

1 _____ of the U.S. Department of Homeland Security released NIMS on March 1, 2004.

    A. Secretary Hillary Clinton
    B. Secretary Tom Ridge
    C. Secretary Loren Hatch
    D. Secretary John Scuddy

2. Some examples of an incident are the following. Select the best response.

    A. Oil and chemical spills
    B. Terrorist/WMDs
    C. Planned events
    D. All of the above

3. The concept of ICS was developed more than 30 years ago, after what event in time.

    A. California Wildfire
    B. Gulf Oil Spill
    C. Fukushima
    D. Chernobyl

4. ICS management functions include all of the following except the following response.

    A. Finance/Administration
    B. Incident Command/Operations
    C. Planning/Logistics
    D. Fire Scope

5. _____ pertains to the number of individuals or resources that one supervisor can manage effectively during emergency response incidents or special events.

    A. Reporting elements
    B. SPAN of control
    C. Minority priority
    D. Unit leader

# IEIA - H-SCADA Certification Series

**LESSON 4: Chapter 6 (pgs 119 -144), Chapter 7 (pgs 145 – 173) and Chapter 8 (pgs 175 – 198)**

This lesson introduces a structural methodology that defines "What to do" scenarios in case of a potential act of terrorism, resulting in a weapon of mass destruction (WMD) deployment, or a hazardous materials spill or contamination. Security Vulnerability Assessments (VA) for any given critical infrastructure or support organization responsible for a critical infrastructure will be addressed. ISO and ANSI guidelines will be addressed for guidelines and standards.

Additional study guide questions and notes are located at the end of each text chapter by the author, Robert Radvanovsky. Chapter 6 (pgs 141-144); Chapter 7 (pgs 169 – 173) and Chapter 8 (pgs 193 – 197).

## Assignment

For Lesson 4, read pages 119 - 198 of your textbook. Complete your reading assignments as outlined below. DO NOT SEND ANSWERS TO IEIA for grading.

## Chapter 6

1. The _____ for the Office for Domestic Preparedness (ODP) is the Department of Justice (now part of the US Department of Homeland Security's Office of State and Local Government Coordination and Preparedness (SLGCP) component responsible for enhancing the capabilities of state and local jurisdiction.

    A. Office of Justice Programs (OJP)
    B. Federal Bureau of Investigations (FBI)
    C. Emergency Preparedness and Readiness (EMR)
    D. Center for Domestic Preparedness (CDP)

2. _____ refers to those individuals who in the early stages of an incident are responsible for the protection and preservation of life, property, evidence and the environment including, emergency response providers as defined within Section 2 of the Homeland Security Act of 2002 (6. U.S.C. 101)

    A. Fire Service
    B. First Responders
    C. HAZMAT Operator
    D. Law Enforcement

3. Awareness level guidelines address training requirements for personnel who are likely to witness or discover an incident or event involving acts of _____ or _____ --- use of WMDs.

    A. Criminal use / terrorism
    B. Public use / safety
    C. Military / private sector
    D. Government / universities

4. Level A: Operations level appear to the _____ level while Level B is the _____ level.

    A. Operations            C. A and B
    B. Technician            D. WMD handler

5. Level A – Training involves possible _____ handling for _____ and other specialized training.

    A. First Responders            C. Hazardous Materials
    B. Weapons of Mass Destruction    D. B and C

6. Level B – know the on-scene situation specific to potential _____ situations or events.

    A. First Responder             C. Hazardous Materials
    B. Weapons of Mass Destruction    D. Chemical Agents

7. When responding to a WMD incident the first responder should know how to select and use the _____ needed to work safely.

    A. Personal Protective Equipment    C. Respiratory System (RS)
    B. Testing Equipment (TE)           D. None of the Above

8. To know the following procedures for performing specialized tasks at the scene of a potential WMD situation or event, the responder should do all of the items listed below. Identify the one that does not apply.

    A. Use technical reference materials
    B. Understand limitations
    C. Follow procedures
    D. Do not assist implant in rehabilitative assessments.

9. First Responders should know how to assess agents or materials used in a potential WMD or hazardous material event based on the sign and symptoms of any individual exposed to the area. They should already know protocols to secure, mitigate and _____ hazardous materials.

   A. Contain
   B. Explodes
   C. Remove
   D. Stage

10. Protective measures involving WMD and hazardous agents or materials should follow post-event rehabilitation best practice for _____.

    A. Assist incident commander
    B. Critical Incident Street Management (CISM).
    C. Be able to assume the terrorist role.
    D. Do not understand the role of the safety engineer.

11. Know appropriate procedures for protecting _____ and minimizing any disturbances of the crime scene to the maximum extent possible, while protecting any individuals at the potential scene.

    A. Terrorist   B. Medications   C. Evidence   D. PPE


**Match the appropriate DOT classifications to the appropriate category. There is only one correct answer.**

_____ 12. Class 1          A. Gases

_____ 13. Class 2          B. Flammable Liquid

_____ 14. Class 3          C. Explosives

_____ 15. Class 4          D. Flammable Solid

_____ 16. Class 5          A. Toxic Materials

_____ 17. Class 6          B. Radioactive Materials

_____ 18. Class 7          C. Corrosive Materials

_____ 19. Class 8          D. Oxidizers

20. Stress effects resulting from hazardous materials incidents can cause increase of a _____ of various stress-related symptoms.

    A. Loss of magnesium
    B. Plethora
    C. Complexity
    D. Fear

## Chapter 7

1. _____ for any given critical infrastructure or support organization responsible for a critical infrastructure.

    A. Unmanned vehicles (UMA)
    B. Risk Assessment
    C. Organizational Plans
    D. Security Vulnerability Assessment (SVA)

2. _____ provides decision makers with information necessary in determining and understanding factors that may negatively influence the operations and outcomes of an organizations operational success.

    A. Identification threats
    B. First Responders
    C. Risk Assessment
    D. Security Agent

3. The American Society of Mechanical Engineers (AMSE) use two basic equations that can determine levels of risk.

$$R_{ai} = F_{ai} \times (vulnerability)_{ij} \times (consequence)_{ij}$$

If $F_{ai}$ is set to "1.0" then the calculated risk is termed a/an _____.

    A. Quantitative Risk Assessment
    B. Unknown Variable Risk
    C. Adversary Attack Asset
    D. Conditional Threat Risk

4. A Quantitative Risk Assessment will apply to all of the measurements stated below, except for one. Identify the one that does not apply.

   A. Determine impacts of information
   B. Determine cost of assets
   C. Analysis is easy to follow
   D. Wide margin of error

5. A Quantitative Risk Assessment is made up of the following measurements of risk except for one of the following listed below.

   A. Difficult to put cost on reputation
   B. Takes subjective issues into account
   C. Pays more attention to information assets
   D. Can reflect personal biases

6. Security Vulnerability Assessment (SVA) is a systematic examination of networks to determine the adequacy of security measurements, identify security deficiencies, provide data from which to predict the effectiveness of proposed security. Measures and confirms the adequacy of such measures after implementation. This includes all systems of assessments listed below but not one. Identify the one that does not apply.

   A. Network – based
   B. Computer – based
   C. Software – based
   D. Non Physical – based

7. A threat affects all of the following except _____.

   A. Intentional or malicious threats
   B. Accident threats
   C. Natural disasters
   D. Does not cause harm to a system

8. _____ is an inherent weakness in a system or its operating environment that may be exploited to cause harm to the system.

   A. Countermeasures
   B. Vulnerability
   C. Framework
   D. Hacker

9. _____ are active processes, procedures and systems features that serve to either detect, deflect, or reduce the probability of a threat, or the impact of vulnerability, there by either reducing or (preferably) removing the system risk.

   A. Vulnerability
   B. VAF
   C. Countermeasures
   D. Systems Analyst

10. The Federal Information System Control Auditing Manual (FISCAM) states that as computers technology advances, government organizations have become dependent on _____.

   A. Tactile Counter Systems
   B. Bio Ethno Life Systems
   C. Computerized Information Systems
   D. Federal Auditory Systems

11. A general control is _____.

   A. Policies and procedures that apply to a large segment of an organizations information system
   B. Ensure transactions are valid
   C. Apply to programmed control technologies
   D. Are appropriately confidential

12. The _____ is the key individual for the audience of an SVA.

   A. Chief Executive Officer
   B. President
   C. Chief Science Officer
   D. Chief Information Officer

13. The initial SVA plan is designed to be used by _____ professionals and auditors.

   A. Safety   B. Security   C. Health   D. HAZMAT

14. _____ are objectives and factors vital to the operational success of the assessment.

   A. Anticipated outcomes
   B. Activities performed
   C. Critical Success Factor (CSF)
   D. VAF or MEIs

## IEIA - H-SCADA Certification Series

Match the following acronyms to the correct response. There is only one correct response.

_____ 15. MEI           A. Vulnerability Assessment Formula

_____ 16. VAF           B. Minimal Essential Infrastructure

_____ 17. CSF           C. Critical Success Factor

_____ 18. ISAC         D. Information Sharing and Analysis Center

19. _____ demonstrates graphically (overall) just how vulnerable an operational system is in terms of its levels of exposure and risks to those threats.

    A. Genetic coding
    B. Chemical coding
    C. Color coding
    D. RNA coding

20. Most assessment color code mechanism weigh the risks (and their threats) using red, yellow and green colors. Match the threat to the color for

    Extremely Risky _____

    A. Yellow     B. Red     C. Blue          D. Green

21. Moderately Risk _____

    A. Yellow     B. Red     C. Blue          D. Green

22. Not Risky at all _____

    A. Yellow     B. Red     C. Blue          D. Green

# IEIA - H-SCADA Certification Series

## Chapter 8

1. The National Fire Prevention Association (NFPA) was established in _____.

   A. 2003      C. 1946
   B. 1896      D. 1776

2. The Security Industry provides range from personal security and safety, to depending how _____ should be configured from a closed – circuit television. (CCTV).

   A. Nano-CMOS     C. MEMS
   B. MITRI         D. Circuitry

3. Employers of state and local governments in states that do not have OSHA-approved health and safety plans must comply with all of the regulations except which one listed below.

   A. EPA 40 CFR 311
   B. SARA Section 126 (8) 40 CFR 311
   C. OSHA 29 CFR 1910.120
   D. NFPA 1600

4. There are _____ different levels of first responders to hazardous material incidents.

   A. 10   B. 4   C. 8   D. 2

5. A voluntary accreditation process for state and local programs responsible for disaster mitigation, preparedness, response and recovery is _____.

   A. Emergency Management Accreditation Program
   B. American Chemical Society Accreditation Program
   C. IEIA Accreditation Program
   D. FEMA Accreditation Program

6. The North American Electric Reliability Council (NERC) represents industry-wide organization in all of the following countries except which one.

   A. United States     C. Mexico
   B. Canada            D. Japan

# IEIA - H-SCADA Certification Series

**Identify the following matches to the appropriate subsection of NERC CIP for proper documents for Cyber Security Standards.**

_____ 7. CIP-002-1      A. Electric Security

_____ 8. CIP-003-1      B. Critical Cyber Assets

_____ 9. CIP-004-1      C. Personnel Training

_____ 10. CIP-005-1      D. Security Management Control

_____ 11. CIP-006-1      A. Systems Security Management

_____ 12. CIP-007-1      B. Recovery Plans

_____ 13. CIP-008-1      C. Physical Security

_____ 14. CIP-009-1      D. Incident Reporting and Response Planning

15. The American Gas Association (AGA) relies on distributed control systems known as _____.

    A. System Control and Data Acquisition (SCADA)
    B. Critical Cyber Assets
    C. Systems Security Management
    D. Recovery Plans

16. IEEE, GTI and AGA all rely on _____.

    A. System Control and Data Acquisition (SCADA)
    B. Critical Cyber Assets
    C. Systems Security Management
    D. Recovery Plans

17. _____ is associated internal, human network and/or machine interface used to provide control, safety and manufacturing operations functionality to continuous batch, discrete and other processes.

    A. ISA-SP99      C. API 1164
    B. AGA 12      D. CIDX

18. _____ standard on control system provides guidance to the operator of petroleum and natural gas pipeline systems.

   A. ISA-SP99
   B. AGA 12
   C. API 1164
   D. CIDX

19. The American Chemical Council's (ACC) Responsible Care Program includes management of practice such as "prioritization" and periodic analysis of potential security threats, vulnerabilities and consequence using acceptable methodologies which is contained in the _____ document.

   A. CIDX
   B. ISO 15408
   C. ISO 17799
   D. AGA 12

20. ISO 15408 is formerly referred to as "Common Criterial" (CC) was initially published as version 2.1 in 1999. This document defines two kinds of documents that are built using a common set of Protection Profiles (PP) and the other is _____.

   A. Targeted Individuals (TIs)
   B. Security Targets (STs)
   C. Trusted Targets (TTs)
   D. Master Organizers (MOs)

21. Health Insurance Portability and Accountability Act (HIPPA) in 1996 was instituted to regulate and protect the confidentiality, integrity and availability of personal health information. It has 4 elements which involve privacy standards. Identify which item does not apply to this regulation.

   A. Privacy Standards
   B. Transaction Code Sets (TCS) standards
   C. Identifiers and Security Standards
   D. Evaluation Study

22. Signed into law in 2005 by President Bush, the Patient Safety and Quality Improvement Act (PSQ1A) of 2005 is intended to encourage the reporting and analysis of _____ by providing peer review protection of information reported to patient safety organizations for the purpose of quality improvement and patient safety.

   A. HIPPA security
   B. Health and Human Accountability
   C. Voluntary Reporting
   D. Medical Errors

23. Gram-Leach-Bliley Act (GLBA) contains provisions to protect consumers personal financial information by _____.

   A. Financial Institutions
   B. Government
   C. Consumer Debts
   D. EPA

24. _____ includes provisions that address audits, financial reporting, and disclosures, conflicts of interest and corporate governance of public companies.

   A. American National Standards Institute (ANSI)
   B. U.S. Department Homeland Security (DHS)
   C. SARBANES-OXLEY Act of 2002 (SOX)
   D. Security Exchange System (SES)

25. FIPS 113 specifies a _____ which, when applied to computer data, automatically and accurately detects unauthorized modification, both intentional and accidental.

   A. Information Technology Management (ITM)
   B. National Security Agency (NSA)
   C. Data Authentication Algorithm (DAA)
   D. Federal Information Security Management Act (FISMA)

## LESSON 5: Chapter 9 (pgs 199 -237); Chapter 10 (pgs 239 – 259) and Chapter 11 (pgs 261 – 287)

Lesson 5 outlines all Information Sharing and Analysis Centers (ISAC) established throughout the US for various critical infrastructure sectors. It will introduce one to terms and concepts of Supervisory Control and Data Acquisition (SCADA) as part of a system and related security methodologies as well as define critical infrastructure information (CII).

Additional study guide questions are at the end of each chapter and additional study notes per author of the text, Roberto Radvanovsky. See Chapter 9 (pgs 230-237); Chapter 10 (pgs 256-259 and Chapter 11 (pgs 279-287).

# IEIA - H-SCADA Certification Series

1. A critical infrastructure asset is an asset (both physical and logical), which is so vital that its disruption, infiltration, incapacitation, destruction or misuse would have a debilitating impact on the health, safety, welfare or economic security of citizens and business. They should include all of the following. Select the best response.

    A. Human
    B. Physical
    C. Cyber
    D. A, B and C

2. Information Sharing and Analysis Center (or ISAC) provides several key services. Identify the one that does not apply.

    A. 24 x 7 (with threat detection)
    B. Members' area
    C. Tailored/customized altering mechanisms
    D. No Risk

3. _____ is a one-stop clearing house for information relating to information technology (IT) threats, physical threats, risks, vulnerabilities and other solutions.

    A. ACS
    B. ISAC
    C. FCC
    D. ATT

4. Surface transportation – ISAL is used for owners and users of the _____ infrastructure(s).

    A. Transportation
    B. Securities
    C. Public Access
    D. GEO-Engineering

5. _____ members include the major freight railroads of the USA, Canada and Mexico as well as Amtrak.

    A. TITC
    B. APTA
    C. AAR
    D. TOP

6. Top Secret and higher clearances are used by _____ for US Military, Academia and IT vendors.

    A. FTA
    B. ITT
    C. EWA
    D. PT-ISAC

## IEIA - H-SCADA Certification Series

7. Water ISAC provides public water systems information from _____ threats to drinking water systems.

   A. Health
   B. Physical
   C. Factories
   D. RAILINC

8. Water Environment Research Foundation (WERF) provides information to _____ and technology to address water quality issues.

   A. Advancing science
   B. Innovation
   C. Environment
   D. Government

9. _____ is the service provider for the Financial Services ISAC (FS-ISAC) under the auspices of the President's Commission of Critical Infrastructure Protection.

   A. Lockheed Martin
   B. Science Application International Corp (SAIC)
   C. Brookhaven Labs
   D. Raytheon

10. _____ is the nation's largest employer-owned research and engineering company.

    A. Lockheed Martin
    B. Science Application International Corp (SAIC)
    C. Brookhaven Labs
    D. Raytheon

11. The North American Electric Reliability Council (NERC) is the _____ that performs the same sector as NERC for physical sabotage and terrorism.

    A. CIPAG
    B. CIP
    C. Electricity Sector (ES-ISAC)
    D. ISAC

12. Identify the organization not included in ES-ISAC and CIPAG co-ordinance.

    A. DOD
    B. API
    C. EEI
    D. IEIA

13. Emergency Management and Response (EMR-ISAC) helps in _____.

    A. NERC Budget
    B. All hazards attack
    C. Independent Board of Trustees
    D. DOE Meetings

14. National Coordinating Center for Telecommunication (NCC-ISAC) provide a _____ , in which the results are sanitized and disseminated in accordance with sharing agreements established for the NCC-ISAC participants.

    A. Congressional review
    B. Network reliability
    C. Library of historical dates
    D. Significant public impact

**Match the acronym with the correct response.**

_____15. Government Emergency Telecommunications Service

_____16. Communicational Resource Information Sharing

_____17. Telecommunications Service Priority

_____18. Shared, Resources High Frequency Radio Program

    A. TSP    B. CRIS    C. SHARES    D. GETS

19. The role of globalization introduces other elements to the _____ for domestic service providers.

    A. National Security Telecommunications Advisory Committee (NSTAC)
    B. Network Reliability and Interoperability Council (NRIC)
    C. NS/EP Communications
    D. Network Information and Exchange

20. Wireless Priority Service (WPS) are authorized and encourages to use _____ to better their probability of completing their NS/EP during periods of wireless and wireline network congestion.

    A. NSC
    B. NS/EP
    C. GETS
    D. WPS

IEIA - H-SCADA Certification Series

21. _____ is the first national program designed by Public Safety for public safety.

   A. PSN
   B. SAFECOM
   C. NCC-ISAC
   D. DHS

22. The American Chemistry Council, FBI and DHS works together to use _____ to establish the chemical sector ISAC.

   A. CHEMTREC
   B. HAZWOPPER
   C. SCADA
   D. HCISAC

23. _____ sector requires a framework through which it can protect the industry from cyber and physical infrastructure threats.

   A. Pet care
   B. Public care
   C. Health care
   D. Home care

24. _____ is a criteria of American Transportation Workers for Highway watch participants.

   A. Transportation of goods
   B. Cargo plans
   C. Anti-terrorist
   D. Road smarts

25. Food and Agriculture ISAC are important because the _____ is a critical national resource.

   A. Food supply
   B. Phone system
   C. Health care
   D. Water

26. World Wide ISAC (WW-ISAC) organizations means of mitigating _____ risks.

   A. Cyber-security
   B. Biotechnology
   C. Biopharma
   D. Educational network

27. Maritime ISAC (M-ISAC) is primary for protecting _____.

   A. Lands
   B. Communications
   C. Harbors
   D. Food

84

## Chapter 10

1. Supervisory Control and Data Acquisition (SCADA) are concerned about computer based _____.

    A. Mechanisms  
    B. Safety  
    C. Health  
    D. Control system

2. _____ is typically used within a single process or generating plant or unutilized over a smaller geographic area or even a single site location.

    A. Distributed Control System (DCS)  
    B. Supervisory Control and Data Acquisition (SCADA)  
    C. Network Demand System (NDS)  
    D. Components of Control Systems (CCS)

3. _____ was used before there was AOL or AT & T and YAHOO/DSL or any popular Internet Service Provider (ISP) they were connected with remote computer systems via modems.

    A. Email  
    B. Specialization  
    C. Wardialing  
    D. BBSS

4. A _____ is an individual who specialized in unauthorized penetration and access of telephone systems.

    A. Mongol  
    B. Phreaker  
    C. Warmonger  
    D. Hacker

5. The main goal of wardialing is _____.

    A. Control  
    B. Access  
    C. Debate  
    D. Systems

6. Wardriving principal and primary goal is access via _____ connectivity.

    A. Wireless  
    B. Land line  
    C. TV  
    D. Radio

# IEIA - H-SCADA Certification Series

7. Warwalking method of wireless network sniffing is performed at (or near) the target point and is performed by a _____ using a PDA device or lap top computer.

    A. Receiver
    B. Banner
    C. Pedestrian
    D. Team

8. Technology research initiatives of control systems have been developed by the following federally funded entities except which one.

    A. Sandia National Laboratories
    B. Idaho National Emergency and Environmental Lab
    C. Los Alamos National Laboratory
    D. Berkley National Labs

9. Control Systems Architecture development for a multi-level network infrastructure will need all of the following except.

    A. Inter-network and interlayer traffic floor network
    B. Fire wall
    C. Intrusion detection systems
    D. Double point systems

10. Businesses need to develop an incident response plan for _____.

    A. Security incidents
    B. Layered security
    C. Segmented networks
    D. Detection systems

## Chapter 11

1. Critical infrastructure information applies to the type of _____ of data information. Example: "Sensitive, But Unclassified."

    A. Network
    B. Designation
    C. System
    D. Group

## IEIA - H-SCADA Certification Series

2. Exemption of the Freedom of Information Act (FOIA) protects _____ owners from disclosure of national security information concerning national defense or foreign policy.

   A. Gifting
   B. Taxes
   C. Disclosure
   D. Crime

3. Exemption 4 of the FOIA exempts from discloser of _____ and commercial or financial information obtained from a person and privileged or confidential.

   A. Records
   B. Trade secret
   C. Reports
   D. Formulas

4. Sensitive but unclassified means some form of protection outside the _____ system.

   A. Formal
   B. Unified
   C. Government
   D. Secret

5. The FOIA allows _____ exemption from this mandatory release policy.

   A. 2    B. 7    C. 9    D. zero

6. _____ is one of the most fundamental security systems.

   A. Systems go
   B. Commercial information
   C. Need to know
   D. Hallway systems

7. The US Department of Energy uses _____ terminology.

   A. Public Access only
   B. For Your Eyes only
   C. Need to Know
   D. "Official Use Only"

8. DOD uses _____ terminology.

   A. Need to Know
   B. For Official Use Only
   C. Official Use – Secret
   D. Unclassified

9. International Traffic in Arms Regualtions (ITARs) prevents forcing entitled from getting notice/security interests of the _____.

   A. UK   B. Canada   C. USA   D. Mexico

10. _____ is a Federal Bureau of Investigation (FBI) program that allows from the information technology indirectly and academia FBI investigations efforts in the cyber area.

   A. Scotch Guard          C. INFRAGUARD
   B. BioShield             D. Non-Internet Public Systems.

~~~~~~~~~~                                              ~~~~~~~~~~

# IEIA - H-SCADA Certification Series

## IEIA - H-SCADA Bio-Energy Field Professional Study Guide for Examination

### MODULE 3

**Advances in Computer: Volume 71 Nanotechnology,
Edited by Marvin V. Zelkowitz**

**TEXT:**
**Advances in Computers: Volume 71 Nanotechnology**
Edited by: Marvin V. Zelkowitz, ISBN: 978-0-12-373746-5
Elsevier/Academic Press, New York, New York © 2007

Module 3 presents the ever-changing landscape in the continuing evolution of the development of the computer within the field of information processing. This module represents the exciting field of nanotechnology. Rapidly moving from the realm of science function to commercial products, nanotechnology involves the creation of microscopic devices that can provide support at a smaller scale previously available. Miniature computing devices may someday provide help, especially in the health and medical fields, which is currently being developed as the 24/7 edible chip to be monitored remotely from your home. This aspect of nanotechnology will address how the devices are made and their integration with the computer, software and individual under SCADA applications.

The text is one that should be studied in its entirety, as this same technology is being applied to environmental monitoring under urban warfare and GEO Engineering applications of telemetry for identification and applications under *First Responder* regulations. The fabrication, characterization and measurement of the nano system will be demonstrated and the subject of its application to artificial and biological entities will be discussed through *bio-scaffolding* applications.

# IEIA - H-SCADA Certification Series

**Assignment**

**LESSON 1**: Read Chapter 1 pages 1-39 of <u>Advances in Computer: Volume 71 Nanotechnology</u>, Edited by Marvin V. Zelkowitz. Complete your reading assignment as outlined below. DO NOT SUBMIT ANSWERS TO IEIA for grading.

1. _____ refers to a set of methods and approaches in physic and chemistry science, engineering fields, biologic and medical areas.

    A. Toxicology
    B. Nanotechnology
    C. Sociology
    D. Singularity

2. In 1959, Richard Feynnon, a future Nobel Laureate gave a visionary talk entitled "There's Plenty of Room at the Bottom." The _____ was developed 40 years later.

    A. Micro procession
    B. Computer
    C. Microelectronics
    D. Silicon chip

3. Molecular technology as applied to nanotechnology in terms of bottom-up technology means the following:

    A. Technology for silver micro process at nano size is assembled from bottom-up.
    B. Size of silicon microprocessor chips
    C. Is not the choice of mass production devices in nanotechnology.
    D. Limitation of top to bottom nano computer components.

4. The computer science aspects of Nanotechnology in research is wide ranging and includes all of the following, except which statement.

    A. Software engineering, networking, internet security
    B. Image processing, virtual reality, intelligent system
    C. Human-machine interface, artificial intelligence
    D. All of the above.

5. F.I. Drexler introduced the term "Nanotechnology" and "Molecular Engineering" in the book, Engines of Creation. He explored many aspects of nanotechnology including potential benefits and _____.

   A. Building products of giant molecules.  B. Decrease in environmental impacts.
   C. Possible dangers to humanity.
   D. Would not benefit humankind as nanomedicine.

6. The first images of nano scale were developed by M. Knoll and E. Ruskee in 1931 using what type of equipment.

   A. Electron microscope           C. High pressure liquid chromatography
   B. GC/Mass spec                  D. Atomic absorption

7. _____ was the first corporation through the work of D.M. Eigler to manipulate individual atoms of Xenon atoms on a nickel plate into the company's name.

   A. GE        C. IBM
   B. SAIC      D. TRW

8. The first step to bottom-up technology is to acquire the ability to manipulate individual atoms at the scale of a nanometer. The people who perform this task are called _____.

   A. Nanoologist    C. Geist
   B. Slavers        D. Nanomanipulators

9. The nanomanipulator was developed at the _____ for multi-disciplinary projects included computer service, physics and chemistry as applied to virtual-reality interface to scanning probe devices.

   A. University of North Carolina     C. Stanford University
   B. University of Central Florida    D. Columbia University

10. Top-down _____ is the process that transfers the geometric shape on a mask to the surface of a silicon wafer by exposure to UV light through lenses.

    A. Krylon Photography       C. Time lapse photography
    B. Photolithography         D. Still photography

11. _____ are crystals that emitted only one wavelength of light when their electron was excited.

   A. Nano rods
   B. Carbon tubes
   C. Quantum dots
   D. Thin films

12. In 1965, G. Moore, the co-founder of Intel, predicted a trend that the number of transistors contained in a micro processor chip would double every 18 months. This became better known as _____.

   A. Murphy's Law
   B. Moore's Law
   C. Law of Nature
   D. Law of Order

13. A _____ became the fifth type of solid state carbon used in bottom-up technology as discovered by S. Iigma of NEC in 1991, after diamond structures, graphite structures, non-crystalline structures and fullerene molecules or Bucky balls.

   A. Carbon nanotube
   B. Quantum dot
   C. Diamond surface
   D. Graphene dust

14. In 1987, T.A. Fulton and G.J. Dolan at Bell Labs developed a single-electron transistor and made observations of the quantum properties. The introduction of the _____ has been able to allow the semiconductor industry to build 70 mega-bite memory chips containing over half-a-billion transistors.

   A. Nano
   B. Nanowire
   C. Nano robot
   D. Nanotube

15. The _____ is DNA computing as developed by L. Adleman in 1994.

   A. Nanocomputer
   B. Nano disk
   C. Nanotube
   D. Nano robot

16. The fact that a _____- molecule can store more information than any conventional memory chip and that _____ can be used to perform parallel computations make the area of nano computing data very appealing.

   A. siRNA ; DNA
   B. SIRT ; RNA
   C. DNA ; DNA
   D. FOX ; SOX

17. Drexler in a book called <u>Nano System</u>, 1992, described _____ - that were biological machines – organic cells – as molecular assembled motor and device.

    A. Nanotubes
    B. Nano radio
    C. Nano computers
    D. Nano robots

18. R.A. Freitas, Jr. in 2003 described a new field of medicine that would be used for its role in gene therapy and diagnostic augmentation.

    A. Neuromedicine
    B. Nanomedicine
    C. Natural medicine
    D. Computerized medicine

19. Nano guided molecular assembling systems, 3-D networking and new system architecture for nano systems, robotics and _____ devices.

    A. Micro molecules
    B. Transitional molecules
    C. Supramolecules
    D. Nanomolecules

20. NASA has been developing a software system, called _____, for investigating fullerene nanotechnology and design molecular machines. It will be used to support and enable the NASA's group to develop complex simulated molecular machines.

    A. Artificial intelligence (AI)
    B. Programmable artificial cell evolution (PACE)
    C. Nano design (ND)
    D. Peptide-DNA

21. Intelligent systems are used in real-world and problems in other research areas. They comprise methods in all of the areas except, which one. Identify one that does not apply.

    A. Algorithms in artificial intelligence (AI)
    B. Nanoscale artificial proto cell, able to self-replicate and evolve under controlled conditions.
    C. Act as nano robots, comprising of an outer membrane, a metabolism and Peptide-DNA to encode information.
    D. They are not being used in micro fluidic FPGA chips.

22. _____ is a software platform for Evolutionary Algorithms (EAs) and ecosystem simulations and for rapid development of telecommunications related applications.

   A. IEIA
   B. EOS (swarm)
   C. ACS
   D. AMA

23. _____ architecture for space exploration by NASA Goddard Space Flight Center is claimed to be a revolutionary mission architecture that was an emergent collaborating behavior of social insects.

   A. Autonomous Nanotechnology Swarm (ANTS)
   B. Evoluable Neural Interface (ENI)
   C. Robot Auto Nano Systems (RATS)\
   D. Perception Particle Swarm Optimization (PPSO)

24. The crucial "_____" in bottom-up nanotechnology is the control of the nano agent.

   A. Missing link
   B. Tiny particles
   C. Complex computers
   D. Swarm systems

25. _____ is inspired by collaborative behaviors in social animals such as birds, ants, fish and termites.

   A. Artificial intelligence (AI)
   B. Degree of intelligence
   C. Self-organization
   D. Swarm intelligence

26. _____ is a chemical that is used in swarm intelligence techniques model the collective behavior in social insects.

   A. Phermone
   B. Bleach
   C. Ammonia
   D. Gasoline

27. _____ is initially used to describe the mechanisms of macroscopic patterns emerging from processes and interaction at microscopic level.

   A. Four concepts
   B. Number of robots
   C. Self-organization
   D. Social insects

28. _____ apply this collective behavior of solve combinational optimization problems and an example is in a traveling salesman problem.

   A. Self-organization
   B. Multiple interaction
   C. PSO algorithm
   D. ACO-meta Heuristic framework

29. Using similar representations of physical agents, the _____ algorithm sums a plausible method for applications including nano robot and coordination control.

   A. ACO
   B. PSO
   C. ITO
   D. CEO

30. PSSO algorithm is designed for optimization problems in physical applications, such as swarm of rescue robots searching for survivors after a/an _____.

   A. Earthquake
   B. Birthday
   C. N-CMOS
   D. Particle swarm

31. Particle Optimization can have additional signals emitted by particles (chemical gradients, electromagnetic programmed fields or different adhesive properties. They are all part the following field/technique _____.

   A. Swarm Optimizaiton
   B. Currier Transfer
   C. Nano Dialing System
   D. Particle Counter

32. _____ (the autonomous construction of a device by itself) is an aspect of robotic engineers.

   A. Payload
   B. Damaged parts
   C. Swarm robotics
   D. Self-assembly

33. Nanorobotics demonstrate the synthetic realization of template self-assembly _____ self-assembly and _____.

   A. BA complexes / chemical molecules
   B. Modular robotics / micro nano scale
   C. Biological / self-reconfiguration
   D. Poly-bot / R. Penrose model

34. What shape best allows a nano self-assembling robot to glide on a cushion of air through programmed electromechanical components.

   A. MTRAN
   B. Hexagon
   C. Triangular
   D. Circular

**Several disciplines of modular robotics produce self-reconfigurable robots using both centralized and decentralized modules. Match the correct type to its description.**

_____ 35. Poly-bot          A. uses cube shaped module with one axis

_____ 36. MTRAN             B. shape of loop, snakes, spider

_____ 37. Poly-bot          C. semi-cylindrical part and 180° axis

_____ 38. MTRA              D. crawler and quadrupled movement

39. For the purposes of creating an artificial self-assembling system, the natural principles of self-assembly can be obstructed by for items. What are they?

   A. Components, environment, assembly protocol, energy
   B. Self-assembling, illustration, mathematics, software
   C. Pivots, joints, levers, steps
   D. Gas, liquid, solid, machine

**Self-assembly illustration can be symmetrical or non-symmetrical. Match the shape with the proper two-dimensional geometric mesocal of self-assembling structures.**

_____ 40. 16-sided polygon    A. Symmetric

_____ 41. Square              B. Non-symmetric

_____ 42. Triangle            C. 3 sets of 2 symmetric shapes

_____ 43. Parallelogram       D. 1 set of 4 symmetric shapes & 3 non-symmetric shapes.

# IEIA - H-SCADA Certification Series

**Assignment**

**LESSON 2:** Nanobiotechnology: An Engineer's Foray into Biology. (Chapter 2, pages 39-102) of <u>Advances in Computer: Volume 71 Nanotechnology,</u> Edited by Marvin V. Zelkowitz. Complete your reading assignment as outlined below. DO NOT SUBMIT ANSWERS TO IEIA for grading.

Nanotechnology is no doubt one of the most significant technologies in the 21$^{st}$ century, which has already brought and will continue bringing profit to our communities. What nano-biotechnology promises is a more significant, deeper and longer-term impact on biomedical and clinical research. We can imagine a world where medical nano devices are routinely implanted or even injected into the bloodstream to monitor health and to automatically participate in repair of the living organism.

1. The _____ structure in the cell nuclei are probed by using protein-sized quantum dots.

    A. DNA  B. RNA  C. Blood  D. Ions

**Nanofabrication form top-down methods use electron beam resists. The resists are made up of hazardous materials that are utilized in lithographic processes. Match the correct (hazardous) resist material to the proper resolution to see as smaller 20 nm as possible. (Note: A through B selections may be used more than once.)**

_____ 2. PMMA (Polymethylmetharylate)     A. Very High

_____ 3. P(MMA-MAA) copolymer            B. Low

_____ 4. NEB-31                           C. High

_____ 5. EBR-9                            D. Very Low

_____ 6. ZEP

_____ 7. UV-5

8. Focused Ion Beam (FIB) nanolithography is used in fabricating silicon _____, bipolar devises and resistor structures and to implant boron and arsenic.

   A. Nano-CMOS    B. MITRI    C. MOSFET    D. AMF

9. Electro Discharge Machining (EDM) is a good example for cutting small or odd-shaped angles, intricate contours or cavities in extremely hard steal on exotic metal. This creates a new _____ system and advanced spark generators have helped with quality machined surfaces.

   A. Bottom and top up (BTU)
   B. Laser beam machining (LBM)
   C. Computer numeric control (CNC)
   D. Ultrasonic machining (USM)

10. A self-assembled mono layer is _____ based.

    A. Chemical lithography
    B. Block polymers
    C. Nanofabrication
    D. Dip-pen

11. Nanophase separation of polymer blends or _____ is a current nanofabrication on self-assembling quantum computing, sensing and integration.

    A. Chemical lithography
    B. Block polymers
    C. Nanofabrication
    D. Dip-pen

12. Nanophase separation using a simple case of block polymers, AB diblocks would use _____.

    A. Gold
    B. Arsenic oxides
    C. Methacrylate
    D. Polystyrene-poly-isoprene (PS-PT)

13. The application of certain cool and heat cycles, periodic pillars arrays used with thin homopolymer film best describes _____, which is a result of a binary system.

    A. Free energy system (FES)
    B. Lithography induced self-assembly (LISA)
    C. PMMA (Polymethylmethacrylate)
    D. (PS-PT) poly styrene-poly isoprene

14. To create one-dimensional standing wave forcing the atomic patterning generated has a _____ detune or _____ detuned.

    A. Blue / Red    B. Yellow / Red    C. Red / Green    D. Blue / Green

15. Complex patterning utilizes _____ atoms which form islands with hexagonal symmetry when using a _____ detuned laser.

    A. Chromium / Red
    B. Iron / Yellow
    C. Cyanide / Blue
    D. Lead / Black

16. _____ et.al. constructed patterns of different letters by passing a beam of ultra-cold metastable neon atoms through a computer-generated hologram that encodes the Fourier transform of a desired atomic pattern.

    A. Einstein    B. Fugitia    C. Tesla    D. Frietas

17. Bottom-Up nanofabrication methods use _____ to crate pattern formation in polymers for this form of film.

    A. Acrylonitirle    B. Siloxanes    C. Polyacetylene    D. Polystyrene

18. Growth of nanotubes and _____ show yet another type of "Bottom-Up" methods of nanofabrication.

    A. Carbon particles    B. Iron oxides    C. Nano robots    D. Nanowires

19. _____ method crates carbon nanotubes through arc-vaporization of two carbon (graphite) rods placed end to end.

    A. Arc-discharge
    B. Chemical vaporaisaiton
    C. Cathode
    D. Anode

20. The best catalyst to use in chemical vapor deposition (CVD) synthesis is _____.

    A. Silver    B. Lead    C. Aluminum    D. Boron

21. _____ refers to a group of replication technologies.

    A. Thermal milling
    B. Nano imprint
    C. Energy dispersion
    D. Scanning electrons

22. _____ nano imprint lithography utilizes UV exposure to harden the resist layer after contact.

    A. Photocurable   B. Strip flash   C. Thermal   D. Laser-assisted

23. _____ is another replication technology that used microfluidic compounds, such as micro mirrors, microgrooves and lens cavities for fiber communications.

    A. Thermal   B. Hot embossing   C. Photocurable   D. Laser-assisted

24. The most widely used nano material is the _____.

    A. Nano particles   B. PT doped Co   C. Magnetic tubes   D. Dipolar wires

25. The use of piezo-driver AFM tip is used to assemble or stack nanoscale objects with high precision. This hybrid approach through nanorobotic manipulation creates _____ that can attain a higher functionality.

    A. Nano Electro Mechanical Systems (NEMS)
    B. PMMA Lithography
    C. Static Connector
    D. Nano-CMOS

26. The use of nanofabrication materials to expand the operator's vision and haptics for a better perception of human machine is part of the emerging field called _____ technology.

    A. SEM   B. AFM (SPM)   C. Virtual Reality   D. Acoustic Band Tools

27. Virtual Reality and haptic systems for the nano world are difficult because of all of the following issues. Identify the one that does not apply.

    A. Time delay and lack of "true" and instant force feed back
    B. Simulation sickness
    C. A and B
    D. None of the above

28. Nanobiosensors are usually coupled to a biological recognition element with a physical _____.

    A. Receiver   B. Stimulator   C. Transducer   D. Repeater

29. The biosensor device does not need physical contact to gain its target to collect biological information and to convert it into forms of _____ that are measurable by means of mechanical, electrical, magnetic, optical and so on approaches.

    A. Time   B. Light   C. Tones   D. Signals

30. During the manipulation and implantation of a biosensor, _____ decreases.

    A. Inflammation   B. Nausea   C. Sleepiness   D. Energy

31. The _____ sensor is affected by temperature, humidity and pH values.

    A. Potassium   B. AFM component   C. Cantilever   D. Piezoelectric

32. _____ nano mechanical sensors utilize ssDNA and SU-8 cantilevers.

    A. Cantilever   B. Polymer   C. Photoresist   D. Amperometric

33. The use of carbon nanotubes, immobilized glucose, hydrogen peroxide and DNA hybridization in the construction of _____ biosensors allow for modification of electrical current between the electrodes and subject biological entities.

    A. Amperometric   B. Polymeric   C. Systematic   D. Biosessmic

34. If DNA is the target and a sensor is used to show a nano switch for the detection of DNA Hybridizaiton this biosensor is called a _____ biosensor.

    A. Atmospheric meter
    B. Hybrid Impedimetric
    C. Conductometric
    D. Potentiometric

35. _____ sensors are used with tunable wavelengths and high quantum efficiency fluorescence, organic dyes, i.e. biological imaging.

   A. Nano-DNA   B. Nanophotonics   C. Nano-CMOS   D. Nano optical

36. The sandwich captured target molecules with a magnetic biobarcode probe is used in bar-code DNA sensors. They are composed of all of the following components except which item.

   A. Magnetic probe
   B. Target molecule
   C. Bar-code probe
   D. Contrast element

37. The following advancements are associated with the capabilities and potentials of biological entities in sensing of environmental parameters are all of the following except, which one.

   A. Stability and degradation
   B. Made of conventional non-biological materials
   C. Harm life
   D. Life time for the environmental biosensor

38. Examples of nanomotors are all of the following except which item.

   A. Protein
   B. Cell based
   C. Artificial respirocyte and microbivore
   D. Nicking enzymes

39. Nanotechnology will be applied to all stages of _____ development, from formulations for optimal delivery to diagnostic applications in clinical trials. And will be based on patents genotype as a factor.

   A. Chemical   B. Enzyme   C. Drug   D. Gene

# IEIA - H-SCADA Certification Series

## Assignment

**LESSON 3**: read Chapter 3, pages 103 – 166): Toward Nanometer-Scale Sensing Systems of <u>Advances in Computer: Volume 71 Nanotechnology</u>, Edited by Marvin V. Zelkowitz. Complete your reading assignment as outlined below. DO NOT SUBMIT ANSWERS TO IEIA for grading.

This lesson will address the development of highly sensitive, selective, reliable and compact sensing systems to detect toxic chemicals and biological agents, which are of great importance to national security.

1. The _____ processing is performed in the pre-brain or the brain.

    A. Olfactory   B. Signal   C. Nasal passage   D. Optical

2. The sponsoring agencies that are interested in the development of an electronic nose are all listed below. There is one listed that does not apply to this interest. Identify the one that does not apply.

    A. National Nanotechnology Initiative (NNI)
    B. Defense Advanced Research Projects Agency (DARPA)
    C. DARPA Mole Sensing Program
    D. EPA and CIA

3. The physiology of the sense of smell is triggered by an odorant molecule that generates an electrical signal. This forms a "_____" by which an odorant may be identified.

    A. Signature   B. Neuron   C. Smell fingerprint   D. Receptor neuron

4. At the heart of the sensing process is the _____ of molecular.

    A. Recognition   B. Bioprocessing   C. Transduction   D. Luminary

5. The odorant receptors are membrane proteins called _____ and derive their name through neuron cell membranes.

    A. Water soluble proteins
    B. AC catalyzes
    C. 7 transmembrane domain G-protein coupled receptors
    D. Dipolarizaiton of $Na^+$, $Ca^+$ and adenyl cyclase (AC)

# IEIA - H-SCADA Certification Series

6. The olfactory bulb has all of the following roles except which one.

    A. Signal Pre-processing
    B. Zones within the epithelium
    C. Temporal aspects of signaling
    D. Does not identify odor

7. Olfactory system as studied in 2002 by Laurent and Perez-Orive confirmed the The convergence of signals in the cortex. They modeled this through the odorant mapped in position to _____.

    A. Height and width   B. Odorant space   C. Vector smells   D. DNA coding

8. Odorant signature is transmitted to other regions of the cortex, this parallel processing includes signature comparison with _____.

    A. Recognition and identity
    B. Stored memories
    C. Olfactory bulb
    D. Transduction flows

9. Electronic Noses for chemical sensing systems need all of following. Identify the statement that is not needed.

    A. Electro Chemical Sensors
    B. Mass-change (piezoelectric) sensors
    C. Optical sensors
    D. None artificial nose systems

10. _____ sensors use a variety of different oxides, most commonly tin, zinc, titanium, tungsten or iridium.

    A. Edible   B. Metal oxide   C. Nano dyes   D. Quantum dots

11. Cyanose® 320 can be customized to measure specific vapors with a hand held device for _____.

    A. 32 polymer carbon black composite sensors.
    B. Personal bandage detectors
    C. Carbon black polymers of methanol and hexane
    D. Tin oxides

12. _____ sensors, the wave travels primary over the surface rather than throughout the device.

    A. Surface acoustical wave
    B. Quartz crystal microbalance
    C. Piezoelectric device
    D. MEMS

13. The use of _____ is/are used in the Array Matric and the Bead Chip sensors as developed by Illumina and partnered with Dow Corning and Chevron.

    A. O Nose™   B. Nanospheres   C. Microbeads   D. Nanotubes

14. _____ nose is used to change color in the presence of the most toxic vapors.

    A. HD printer
    B. Colormetric
    C. Pixels
    D. White color

Match the acronym with its proper definition.

_____ 15. SWNT          A. Multiwalled nano tubes

_____ 16. Nanobelt      B. Tin oxide that service CO, NO2 with a charge

_____ 17. MWNT          C. Tin oxide sense N2 and CO with conductance charge

_____ 18. Nanowire      D. Single walled nano tube

19. Leiber at Harvard and Moskovits at UC Santa Barbara have demonstrated that nanowires can behave like _____.

    A. Genetic molecules
    B. Sensors
    C. Small molecules
    D. Tools

20. The design for an electronic nose system integrated on the nanometer scale consists of all of the following except which item.

   A. Nanowire sensing array
   B. Nano memory sensing system
   C. None of the elements
   D. Nanowire cross bar array

**Match the correct Acronyms with the correct definition. See pages 158 and 159 of text.**

_____ 21. ANN          A. Volatile organic compound

_____ 22. VLSI         B. Olfactory bulb

_____ 23. VOC          C. Artificial neural network

_____ 24. OB           D. Very large scale integrated

_____ 25. DFA          A. G protein specific to the olfactory system

_____ 26. BL           B. Discriminant function analysis

_____ 27. RDA          C. Karlsrube Mikro-Nase (Kerlsrube Research Center ) Micro Nose

_____ 28. Golf

_____ 29. KAMINA       D. Advanced research and development activity

_____ 30. PLL          A. Phase-locked loop

                       B. Barley lectin

                       C. Quartz crystal microbalance

                       D. Surface acoustic wave

### IEIA - H-SCADA Certification Series

**Assignment**

**LESSON 4:** Chapter 4, pages 167 – 251: Simulatin of Nanoscale Elecronic Systems of <u>Advances in Computer: Volume 71 Nanotechnology</u>, Edited by Marvin V. Zelkowitz. Complete your reading assignment as outlined below. DO NOT SUBMIT ANSWERS TO IEIA for grading.

As modern silicon integrated devices have reached nanometer scale, device simulation has evolved to provide designer's with physical tools that account for particle and quantum function of transport. This lesson will present an overview of simulation approaches for nanoscale device systems with applications examples ranging from traditional, MOSFETS to molecular and bio-inspired structures.

1. Metal-oxide semiconductor filed effect transistor (MOSFETS) are composed of integrated circuits that are sandwiched between a layer of _____.

    A. Arsenic and gold
    B. Silicon dioxide
    C. Lead
    D. SOD – superoxide dismutase

2. The use of AlGaAs in increased orders of magnitude have led to the erection of new materials _____ for ballistic approaches.

    A. GaAs alloids
    B. Nano-CMOS
    C. Nano-MOSFETS
    D. High-electron mobility transistors (HEMT)

3. Silicon technology has been the _____ architecture based on a basic inverter structure which only conducts current when switched between two logic states takes place. (It also has a nano version).

    A. C-MOS     B. MITRI     C. HEMT     D. MEMS

4. _____ are related to reliability problems when they reach an energy threshold.

    A. Hot carrier   B. Trouble maker   C. First responder   D. Two gater

5. Quantum approaches to the complete hierarchy approaches to make device simulations, involves the _____ equation and _____ particle approach.

   A. Boltzmann
   B. Monte Carlo
   C. Drift Diffusion
   D. A and B

6. In the _____ approach, carrier transport is viewed as the transmission and reflection of carrier fluxes with semiconductor for thin slice (1 D) meshes (2D), so that these regions are sufficiently small to ascended constant doping and fields within.

   A. Scattering matrix
   B. Monte Carlo approach
   C. Boltzmann
   D. All systems go

7. Advanced MOSFET devices have approached a region where transport consistence of an admixture of all of the following features, except which one.

   A. Ballistic   B. Quantum   C. Semi-Classical feature   D. Hydrodynamic

8. The replacement of transistors by _____ is appealing for scale, since the atoms involve a volume of $10^6$ to $10^9$ atoms.

   A. Sand   B. Metal   C. Molecules   D. Neutrons

9. Carbon nanotubes (CNTs) can form a _____ lattice that makes an armchair or zig zag effect.

   A. Graphene   B. Silicon   C. Siloxane   D. Non-carbon

10. Biological systems affect a base of electrolyte fluid flow in vessels and _____ that involve a wide spectrum of length and time scales.

    A. Amino acids   B. Bionutroic devices   C. Ion channels   D. pH gradient

**Match the proper space scales in biological processes to the proper physical feature and/or application.**

_____ 11. Electronic structure of molecules

_____ 12. Conformational searches, molecular stimulation and molecular Electrostatics

_____ 13. Brownian dynamics

_____ 14. Diffusion, migration and fluid flow

    A. Atomic and subatomic
    B. Molecules
    C. Mesoscopic
    D. Continuum

15. The _____ is the web portal of the Network for Computational Nanotechnology as performed with code bio-MOCA as developed by the University of Illinois.

    A. Nano Hub   B. 2 GHz   C. Xeon   D. Monte Carlo

**Assignment**

**LESSON 5:** read Chapter 5 pages 251 – 271, Identifying Nanotechnology in Society of <u>Advances in Computer: Volume 71 Nanotechnology,</u> Edited by Marvin V. Zelkowitz. Complete your reading assignment as outlined below. DO NOT SUBMIT ANSWERS TO IEIA for grading.

Manufacturing materials and systems with compounds thousands of time smaller than the width of a human hair promises vast and sometimes unimaginable advances in technology. This lesson will address what different groups are referring to when they say nanotechnology, how this relates to the science involved and how the various definitions of this broad field of endeavor might be improved.

## IEIA - H-SCADA Certification Series

1. The societal implications of nanotechnology accounts for nearly ___ of direct federal funding in the USA.

    A. Zero    B. 10%    C. 80 %    D. 100%

2. A person who studies nanotechnology is called a _____.

    A. Toxicologist
    B. Environmental Engineer
    C. Computer Engineer
    D. Nanotechnologist

3. Fear as much as hope has popularized nanotechnology, from the hysterical vision of "grey goo" to the quite realistic fears of _____.

    A. Death
    B. Cancer therapy
    C. Human toxicity
    D. Biometrics

4. The futurist concern of nanotechnology is concerned with _____ and the fiascoes of genetically modified organisms (GMO) in food.

    A. Semiconductors            C. Nano machines
    B. Sensors                   D. $H_1N_1$ virus

5. The _____ has compiled a database of all products that claim to use nanotechnology. But these companies that are actually using the nanotechnology today – nano particles – in situations that may be hazardous (such as cosmetics) do not advertize this fact.

    A. Ronald Regan Library
    B. George W. Bush Library
    C. Woodrow Wilson Center
    D. Smithsonian Insitute

6. The strongest shaper of nanotechnology has been the U.S. _____.

    A. Presidents                C. Military
    B. Federal government        D. Intelligence community

7. Nanotechnology in a 2006 CIRCA had two critical views in the Yin and Yang symbolism which was the _____ and the reality.

    A. Vision    B. Gift    C. Horror    D. Manipulation

8. The _____ defines nanotechnology as research and technology or development of products regulated by the agency as to the ability to control or manipulate the product on the atomic scale.

    A. EPA    B. DHS    C. OSHA    D. FDA

9. The smaller the molecules (amount of atoms) to the _____ temperature, the properties which occurs at the nano scale is referred to as "quantum size effect."

    A. Color    B. Hot    C. Cold    D. Numbing

10. Nanowires are also classified as _____ since they are solid objects as opposed to hollow nanotubes.

    A. Nano particles    B. Nano crystals    C. Mesogens    D. Bioscaffolds

11. _____ is when you sandwich a thin sensor conductor layer between two bigger semiconductor layers of a different type.

    A. Quantum dot    B. Quantum particles    C. Nano particle    D. Quantumwell

**Assignment**

**LESSON 6:** Read Chapter 6 pages 273 – 296 of <u>Advances in Computer: Volume 71 Nanotechnology</u>, Edited by Marvin V. Zelkowitz. Complete your reading assignment as outlined below. DO NOT SUBMIT ANSWERS TO IEIA for grading.

The prominence of nanotechnology as a matter of national policy is significant, as is the attention being afforded to ethical and societal consideration. This lesson will address the ethical policies of nanotechnology as applied to society, scientists, government, industry and military.

1. Nanotechnology global funding is both _____ and _____.

A. Elusive and contentious     C. Presidential and mandatory
B. Responsible and research    D. Funding and raw

2. The fundamental social economic ethical frame work is _____ for nanotechnology.

   A. Significant challenges           C. Rule based reasoning
   B. Influence the direction of research    D. Bends the opposition

3. The NSF will have funded nanotechnology in the USA with $1 trillion by year 2015, so that the crucial field of nanotechnology would avoid economic and _____ disastrous consequences.

   A. DHS    B. Militarily    C. Environmentally    D. Health wise

4. The ETC report contained recommendations by the Prince of Wales and his concern for _____.

   A. Grey goo    B. GMOs    C. Crichton's Prey    D. MNTs

5. U.S. President _____ signed the 21$^{st}$ Century Nanotechnology Research and Development of 2009 public law (97, PL 108-153), which was to accelerate the development and applications of nanotechnology into the _____.

   A. Military           C. Public sector
   B. Private sector     D. Academia

6. The 21$^{st}$ Century Nanotechnology Research and Development of 2003 aimed at more clearly demonstrated norms such as avoiding plagiarism and ethical dilemmas such as _____.

   A. Proper closure      C. Professional conduct
   B. Whistle blowing     D. Coupling of scientific research

7. Collaboration tasks for nanotechnology include all of the following except one of the following items.

   A. Research              B. Social-technical inquiry and interaction
   C. Responsibility and reputation
   D. Non-participation of self-critical federal science

# IEIA - H-SCADA Certification Series

## IEIA - H-SCADA Bio-Energy Field Professional Study Guide for Examination

### MODULE 4

### Design for Manufacturing and Yield for Nano-Scale CMOS
### By Charles C. Chiang and Jamil Kawa

**TEXT:**
**Design for Manufacturability and Yield for Nano-Scale CMOS**
Charles C. Chiang and Jamil Kawa, ISBN: 978-1-4020-5187-6
Springer, Netherlands © 2007

Module 4 will walk an individual through all the aspects of manufacturability and yield in a nano-CMOS process and how to address each aspect at the proper design step starting with the design and layout of standard cells and how to yield grade libraries for critical are and lithography artifacts through place and route, CMP-model based simulation and dummy fill insertion, mask planning, simulation and manufacturing and through statistical design and statistical timing closure of the design. It alerts the designer to the pit falls to watch for and to the goo9d practice that can enhance a design's manufacturability and yield. It will enable the examinee to practice IC designs and EDA tool for development.

### Assignment

**LESSON 1:** Read Chapters 1, 2, 3 and 4 (pages 1 – 168) and include Preface pages XX1- XXV of the text: <u>Design for Manufacturability and Yield for Nano-Scale CMOS</u> by Charles Chiang and Jamil Kawa. Springer. Netherlands © 2007

1. _____ industry is the manufacturing process based on optical lithography.

   A. Applied Sciences (AS)
   B. Design for Yield (DFY)
   C. Integrated Circuits (IC)
   D. Critical Area Analysis (CAA)

2. The following IC process follows Moore's Law, solution to challenge of high significance for the following manufacturing steps except which one.

   A. Illumination sources of light with shorter wave length
   B. Mask writer (E-BEAMS)
   C. Reticules with fewer defects
   D. Uses ammonia compounds capable of higher current densities.

3. One change in a specific element leads to better ICs. What is the element?

   A. Copper    B. Aluminum    C. Zinc    D. Potassium

4. Optically an IC needs a light source with a wavelength of __ nanometers (nm) with a big gap extending all the way to the _____ nanometer ultra-violet (EUV) light source.

   A. 193    B. 280    C. Zero    D. 1

5. The use of intra-metal oxides to reduce capacitance (improve speed) encountered big challenges in how to reduce _____ dielectric materials without increasing leakage.

   A. Mg magnesium    B. K potassium    C. Na sodium    D. Cl chloride

6. Systematic variations in ICs impact all of the following except which one of the following selections.

A. Leakage    B. Time    C. Yield    D. Hand-you

7. The methodology of ensuring that a product can be manufactured repeatedly, consistently, reliability and cost effectively is called _____.

   A. Design for manufacturability (DFM)
   B. Circuit metal oxide systems (CMOS)
   C. Integrated Circuits (IC)
   D. Applied Services Networks (ASN)

8. The nano era of _____ aided the IC industry.

   A. IC    B. DFM    C. CMOS    D. ASN

9. _____ migration (including open vias) was the main problem with aluminum (Al) as compared to copper.

   A. Ammonia   B. Metal   C. Silicon   D. Thiophenes

10. _____ scaling has resulted in gate oxide thickness reaching the limit of a few mono-atomic layers that is very hard to control.

    A. Copper deposition   B. Low K   C. High K dielectrics   D. Gated oxide

11. The native silicon has been used for enhancing the mobility of p-carriers or n-carriers involving the intentional creation of mechanical stress. This is caused by the following materials except which one.

    A. C   B. SiN   C. SiC   D. SiGe

12. A light source candidate for a lithography light source has been _____ 193 nm Light source inlet extreme ultra violet (EUV).

    A. ArF   B. SiC   C. SiO2   D. Ne

13. _____ is defined as the set of optical and geometrical (layout) procedures performed individually or in any particular combination to enhance the printability of a design feature and to meet design intent.

    A. SRAF   B. PSM   C. RET   D. OPC

14. _____ delivers a technology node performer advantage at the same power level as an existing _____ technology mode, or delivers the same performance at a significantly reduced power 15 to 20 percent.

    A. CMOS/FET
    B. SOI/CMOS
    C. FinFet/T
    D. MEMs/CNTs

IEIA - H-SCADA Certification Series

Match the following acronyms with the proper name.

_____ 15. MEMS        A. Silicon on insulator

_____ 16. SWCNT       B. Single walled carbon nanotube

_____ 17. SOI         C. Chemical mechanical polishing

_____ 18. CMP         D. Micro electronic material sensors

19. The cost of a silicon re-spin is 6 months and from $25 million to $40 million. By the year 2009 the node mask preparation file data size will be 45 nm and _____ Giga bites.

   A. 144    B. 486    C. 1,094    D. 65

20. An extended full system on chip (SOC) design will include all of the following. Select the most comprehensive response.

   A. Logic systems, floor planning
   B. Mask systems
   C. Placement
   D. A, B and C

21. In the _____ method, the name is self-evident in terms of the fact that "expanding" the circle needed to cause a short or an open define the particle size that causes such a short or open circuit.

   A. Poisson base yield.
   B. Intre-layers
   C. Shape expansion
   D. Open CA on a wire

22. A cell library yield grading starts at the very basic steps of the design. It consists of a _____ number and metal.

   A. Poly CA    B. ASIC    C. Mi    D. Metal 1

23. _____ are all of the movable wires in the layout that are computed in one sweep through the layout.

   A. Space-visible neighbors
   B. Wire pushing
   C. Algorithm
   D. DRC

24. _____ is the original wire that is replaced by a set of horizontal and vertical wire segments such that most of the original wire is now located at the calculated optimal position while obeying the spacing rules of the given layer.

   A. Wire pushing
   B. Space-visible neighbors
   C. DRC
   D. Alogorithym

25. The key benefits of the proposed algorithm uses the _____ and a _____ to decide which wire to widen and which ones to spread.

   A. Layer CA / wire CA
   B. Space visible wire/ neighbor
   C. Original wire/ wire
   D. Analog/ CA wire

26. Argon Fluoride (ArF) illumination sources are used in nano-CMOS for _____ Lithography.

   A. Gamma half life
   B. Systematic yield
   C. Cosmic bursts
   D. IC failure

27. _____ has a gamma wavelength of 248 nm and is used in nodes 130 nm and higher.

   A. Argon fluoride   B. Krypton fluoride   C. Super neon   D. Teflon coatings

28. _____ are coated with a photo resist (positive and negative) and the projected light develops the photo resist such that the projected pattern (or its complement) is etched way with chemicals defining the pattern on the silicon wafer.

   A. Cookie   B. Cupcake   C. Wafer   D. Pancake

29. Optical System Interaction uses corner rounding to get exacerbated phenomena in _____ lithography.

   A. CD under forcer
   B. 3-D fluctuations
   C. Mask writing
   D. Sub-wave length

30. Sub resolution assist feature (SRAF) are _____ placed on the mask to enhance the image of adjacent features.

   A. Large dots   B. Medium circles   C. Small lines   D. Diamond spheres

31. Wafers are coated with photo resist, which is a _____ sensitive to light.

   A. Polymer substance
   B. Toxic chemical
   C. Plastic makers
   D. Dye enhancer

32. What metal is opaque and blocks light as a phase shift mask (PSM)?

   A. Silicon   B. Chrome   C. Oxygen   D. Soap

33. _____ is the third work horse of the nano era lithography besides OPC and PSM.

   A. Depth of focus
   B. Molybdenum silicide
   C. Off axis illumination
   D. Attenuated PSM (APSM) (AUPSM)

# IEIA - H-SCADA Certification Series

34. _____ and _____ have been very useful in 65nm technology of pupil alternatives.

   A. Conventional/ annular
   B. Dipole/ ASML dipole
   C. Quadruple/ annular
   D. Cross/ASML Quasar

35. The off axis illumination optimize DOF for a particular _____.

   A. Pitch   B. Harmonic   C. Phonon   D. Light photon

36. Chrome-less Phase lithography (CPL) is a _____, single mask technology.

   A. Single exposure
   B. Rapid rate
   C. Gas guzzler
   D. Synopsys ICC

37. A _____ is an example of a uni-directional overlap for a vertical slicing direction.

   A. Rocket   B. Stair case   C. Ladder   D. Chair

38. A _____ is an example of a bi-directional overlap for a vertical slicing direction.

   A. Rocket   B. Staircase   C. Ladder   D. Chair

39. The _____ is characterized by very high costs, paradigm shift, sub-wave length and advanced mask.

   A. Nano CMOS   B. MOFFET   C. ICP   D. DOF

# IEIA - H-SCADA Certification Series

**Assignment**

**LESSON 2:** Read Chapters 5, 6, 7 & 8 (pages 99 – 242)) of the text: <u>Design for Manufacturability and Yield for Nano-Scale CMOS</u> by Charles Chiang and Jamil Kawa. Springer. Netherlands © 2007

Lesson 2 will cover the important planarization procedure of chemical mechanical polishing (CMP), wafer level variability parametric yield, static tiring, critical scale method and other related topics.

1. The primary metal alloy for IC manufacturing is all of the following except which one.

    A. Gold (Au)   B. Silver (Ag)   C. Copper (Cu)   D. Zinc (Zn)

2. _____ was the original standard metal used in the IC manufacturing.

    A. Aluminum (Al)   B. Copper (Cu)   C. Silver (Ag)   D. Gold (Au)

3. In dielectric systems of ILD the uniform layer has 1 layer of a 90 nm node and 3 layer sandwich with the top and bottom layer being 5000 Angstrom of _____.

    A. SiO2   B. AlCl   C. K   D. FeO2

4. Barrier films for _____ are evaluated in terms of their ability to block all Cu diffusion at 800 degrees C.

    A. Copper (Cu)   B. Zinc (zn)   C. Gold (Au)   D. Silver (Ag)

5. There are three kinds of Post ECP Topographies of a wire. Identify the one that is not a Post ECP.

    A. Conformal fill
    B. Super fill
    C. Over fill
    D. Land fill

6. ILD CMP Model based _____ is complicated process used in Nano-CMOS.

    A. No fill            C. Ice filling

# IEIA - H-SCADA Certification Series

   B. Smart dummy filling   D. CMP planarization
7. A scan and step lithography system represents a _____ moved to the right.

   A. Optics   B. Light source   C. Mask   D. Wafer

8. Environmental variability and aging are characteristics of chip(s) used for temperature and _____.

   A. Voltage   B. Home   C. Electricity   D. Smart meters

9. Some SSTA issues, concerns and approaches in a critical path method CPM include all of the following except which one does not apply.

   A. Simplicity   B. Tractability   C. Conservatism   D. Loners

10. The process of _____ improves the open CA yield number, again as long as it is not done at the expense of the shorts CA number for OPC compliance routing.

    A. Wire widening   B. Smart metal fill   C. Wire-spreading   D. Boundary

# IEIA - H-SCADA Certification Series

## IEIA - H-SCADA Bio-Energy Field Professional Study Guide for Examination

### MODULE 5

### FCC Online Table of Frequency Allocations (47 C.F.R. Statute 2.106) by FCC Office of Engineering and Technology
### Revised on May 25, 2012

**TEXT:**

**FCC Online Table of Frequency Allocations (47 C.F.R. Statue 2.106)**
Federal Communications Commission Office of Engineering and Technology
Policy and Rules Division, Revised on May 25, 2012
US Governmental Printing Office, Washington, DC

***Disclaimer: The Table of Frequency Allocations as published by the federal Register remains the legal source document.***

Module 5 will concentrate on the physical use of the FCC Online Table of Frequency Allocations, which will allow a SCADA operator to determine the frequency being emitted to the targeted source. This will allow an individual to become familiar with looking up specific frequencies that have been identified as foreign bodies (sensors within a biological system) and/or foreign sources that are targeting a specific location. The more you use the document the more you will become familiar with its various sections and allocations. The allocations will also identify its signal source, intent and source/country of origin.

# IEIA - H-SCADA Certification Series

**Assignment**

**LESSON 1:** Familiarize yourself with the proper frequency allocations as published by the FCC in the Federal Register. See pages 1 - 97

**Match the correct response to the proper region and or frequency range.**

_____1. Region I: 9 14          A. Radionavigational

_____2. 72 - 84                 B. Maritime Mobile

_____3. 20.05 - 59              C. Maritime radio navigational

_____4. 90                      D. Private land mobile

_____5. 74.8 - 75.2             A. Land mobile

_____6. 54 - 72                 B. Aeronautical Radio Navigational

_____7. 47 - 49.6               C. Radio Astronomy

_____8. 73 - 74.6               D. Broadcasting

_____9. 130 - 134               A. Amateur Radio

_____10. 97                     B. Earth Exploration Satellite

_____11. 123 - 130              C. Fixed Satellite (space to earth)

_____12. 148.5 - 151.5          D. Earth Exploration (passive) Radio Astronomy Space Research (passive)

# IEIA - H-SCADA Certification Series

**Assignment**

**LESSON 2:** Frequency code familiarization exercise utilizing type and allocations of FCC document pages 68 – 168.

**Match the frequency to the correct response.**

_____ 1.  70 – 72 Khz                A. Bangladesh and Pakistan

_____ 2.  525 – 535 Khz              B. Region 2

_____ 3.  450 – 470 MHz              C. IMT

_____ 4.  1525 – 1530 MHz            D. Saudi Arabia

_____ 5.  10.7 – 11.7 GHz            A. 5.407 Argentina

_____ 6.  2310 – 2360 MHz            B. Canada, acoustical mobile

_____ 7.  2500 – 2520 MHz            C. Region 1 fixed Satellite

_____ 8.  2360 – 2400 MHz            D. 5.393, US, Canada, India, Mexico

_____ 9.  US 222                     A. GOES

_____ 10. US 226                     B. VHF

_____ 11. US 342                     C. 13360 – 13410 MHz

_____ 12. US 343                     D. DGPS

# IEIA - H-SCADA Certification Series

## IEIA - H-SCADA Bio-Energy Field Professional Study Guide for Examination

### MODULE 6

**FDA/OSEL Annual Report 2007 by Food and Drug Administration Center for Devices and Radiological Health**

**TEXT:**
**FDA/OSEL Annual Report 2007**
Food and Drug Administration's Center for Devices and Radiological Health (CDRH) and Office of Science and Engineering Laboratories (OSEL)
Larry G. Kessler, Sc.D., Director OSEL/DCRH & Chair, Global Harmonization Task Force

Module 6 addresses the regulatory and research support of the FDA in various aspects of health with specific concern for telemetry and sensors that are used in food, animals and medicine (humans). This following exercise is to familiarize you with this aspect of the FDA that also addresses its interface into SCADA.

# IEIA - H-SCADA Certification Series

**LESSON 1 and LESSON 2:** Read pages 1 – 73 the Office of Science and Engineering Laboratories 2007 Highlights for Laboratory and their augmentation through OSEL.

Match the Office with the research activity as applied to nanotechnology and/or Toxicological Risk Assessments for EMI's. There will only be one correct answer per response. DO NOT SUBMIT YOUR ANSWERS TO IEIA.

_____ 1. Silver nanoparticles             A. National Toxicology Program

_____ 2. iPods                            B. LCIT

_____ 3. Cardiovascular Therapies Swine mode  C. ANSI/AAMI 69.2000

_____ 4. Light tissue extraction          D. Optical Diagnosis Lab

_____ 5. National Medical Device Regualtory   A. Optical and Medical Nanophotonics

_____ 6. Low level light therapeutics     B. MATLAB

_____ 7. Software                         C. GHTF

_____ 8. TiO$_2$/ cytokines (J774)        D. Toxicology

**IEIA - H-SCADA Bio-Energy Field Professional
Study Guide for Examination**

**MODULE 7**

**Guidelines on the Evaluation of Vector Network Analyzers (VNA)
Euramet
By European Association of National Meterology Institutes.
EURAMET cg-12**

**TEXT:**

**Guidelines on the Evaluation of Vector Network Analyzers (VNA)
EURAMET (European Association of National Metrology Institutes)**
EURAMET cg-2, Version 2.0 (03/2011) Previously EA-10/12
Calibration Guide, Braunschweig, Germany  e-mail:
secretariat@euramet.org   Phone:  +49 531 592 1960

Module 7 addresses the guidance on measurement practices in the specified fields of measurements. By applying the recommendations presented in this exercise one will produce calibration results that can be recognized and accepted throughout Europe and the United States. Many of the equipment used in the field of identifying SCADA systems will follow some form of calibration. It is necessary to familiarize yourself with other areas of calibration, which will allow the participant to be more diverse in understanding comprehensive data collection and computer network integrated systems of simple tasks.

# IEIA - H-SCADA Certification Series

**Assignment:** Read pages 1 – 19

**LESSON 1:** Covers the guidelines on the Evaluation of Vector Network Analyzers (VNA)

**LESSON 2:** Discusses the uncertainty of evaluation for on-port measurements (computer based) and UVRC.

1. The VNA document gives guidance on measurement practices in the specified fields of _____.

   A. Time tables   B. Measurements   C. Geometry   D. Distance

2. The document has been used by third party _____, peer reviewers, witnesses to measurements and many other professional areas of expertise.

   A. National Accreditation Boards
   B. IEIA and ACS
   C. OSHA
   D. NAS

3. The line diameter for the IEC Standard 60457 and IEEE Standard 287 [2] states that a GR 900 or equivalent connector must match a _____ line diameter.

   A. 7 mm   B. 3.5 mm   C. Type-N   D. 14 mm

4. The _____ should conform to the normal requirements of accreditation with respect to equipment, reference standards, operating instructions, method calibration, uncertainty condition and other precautions.

   A. Cafeteria   B. Laboratory   C. Fields   D. Space

5. A _____ kit has the purpose to demonstrate traceability to national standards on a contracting basis.

   A. Accreditation body
   B. VNA standards
   C. Traceability kit
   D. Manufacturers cost

6. The vector network (VNA) analyzer should have different _____ models and associated calibration measurements to meet the VNA's operational specifications.

   A. Mathematical   B. Scientific   C. Environmental   D. Computer

7. At low frequency about below 500 MHz the available online became less ideal for _____ Evaluation.

   A. VNA   B. Philosophy   C. Black box   D. Uncertainty

8. System repeatability (Resolution and Noise); connector repeatability; effects of cable flexure and effects of ambient conditions are all a part of uncertainty evaluation(s) from _____ measurements, UVRC.

   A. Two port   B. 10 port   C. One port   D. No ports

9. The equation below

   $$M = \frac{\text{Maximum Ripple Amplitude}}{2}$$

   is used as a function of _____ from the magnitude of the ripple.

   A. Tone   B. Frequency   C. Gas   D. Light

10. The relative uncertainty in tracking reflection measurements are _____ as half interval of a rectangular distribution.

    A. 0.001   B. Zero   C. 10   D. 100

11. To measure VNA linearity for accredited range is _____ dB for the range of 0 to 80 dB.

    A. 0 to 70   B. 10   C. 0.3 to 10   D. 9

12. Isolation after calibration is known as _____.

    A. Cross-talk   B. Crackers   C. Lime based   D. A soaker

13. The set of the two data checks of VNA (one and two ports) are the two sets of data; should differ by more than the combined, _____ of uncertainty then further investigation is necessary.

    A. Right angle
    B. Root sum of square
    C. Joint apex
    D. Zero vector

# IEIA - H-SCADA Certification Series

## IEIA - H-SCADA Bio-Energy Field Professional Study Guide for Examination

### MODULE 8

Telemedicine and e-Health: Abstracts from the American Telemetric Association

**TEXT:**

**Telemedicine and e-Health: Abstracts from the American Telemedicine Association**
Eighteenth Annual International Meeting and Exposition,
May 5-7, 2013 – Austin, Texas  ISSN 1530-5627  Mary Ann Liebert, Inc. Publishers

Module 8 is to familiarize you with the use of telemetrics in medicine and its overlap into the new field of e-Health. Many of the new nanotechnology delivery systems and biosensors are designed to stay so long in a body or even to monitor various parameters. It is advised to familiarize you with the terms and types of research and concerns that are being addressed as this area of medicine will transpire into a potential environmental concern and an applied application to existing or new SCADA systems.

**LESSON 1:** Will address the foundation for starting a Telemedicine Program within your facility.

**LESSON 2:** Will evaluate methods concepts and next wave – competing.

Remember this document is to allow the reader to familiarize themselves with the technology, topics of and its overall use in the medical community.

1. Dr. Rashid Bashshur presented a paper on the impact of telehealth on _____ reduction.

A. Cost   B. Time   C. Health   D. Diagnostics

2. For a successful telemedicine business enterprise you must have _____ and operation skills.

   A. Libraries   B. Data   C. Finance   D. Tools

3. It is important to consider the _____ design for the development of a large-scale telehealth program.

   A. Prosecutor
   B. User-centered
   C. Experience
   D. Communication

4. To deliver specialized health care services and technologies _____ may be used.

   A. Notebooks   B. Cookies   C. Mobile Apps   D. Software

5. In section # 006 on May 6, 2013 the use of telemedicine was identified for _____ practice guidelines and pearls.

   A. Teleneurology
   B. GEO Engineering
   C. Surgical Studies
   D. Teledermatology

6. Remote neurocognitive assessment on military and civil patients may be performed by the _____.

   A. Staff   B. Neuro Department of Telehealth   C. Pathology   D. Safety

7. Telemedicine may be used in multi-state health as presented by _____.

   A. Xulan Wang
   B. Gary Capistrant
   C. April M. Armstrong
   D. Munro Cullum

8. Telehealth may be used for all of the following except one. Identify the one that does not apply.

   A. Creating better disease management
   B. Neurology
   C. Members only
   D. Nimbility via integration

9. The _____ Olympics use telemedicine venues.

   A. 1937 German    B. 2012 London    C. 2016 Japan    D. 2006 US

10. Telemedicine may be used for health and _____ aid as presented by Peter Killecommon.

    A. Communication   B. Chasing   C. Reimbursement   D. Humanitarian

# IEIA - H-SCADA Certification Series

## IEIA - H-SCADA Bio-Energy Field Professional Study Guide for Examination

### MODULE 9

### Energy Field, Radiation, Health and Safety Issue

**TEXT:**

**Section 8: Health Effects Associated with Exposure of Industrial Workers to Radiofrequency Waves**
RF Toolkit-BCCDC/NCCEH pages 191 – 216, CDC, Atlanta, GA

**Document:** SCAN TECH: EMF Levels and Safety A Comparative Guide, pages 1 – 8. www.scantech7.com  Scan Tech 7405 Wild Valley Drive, Dallas, Texas Phone: 214-912-4691 used with permission for IEIA SCADA Reference Library.

It can be very hard to say exactly what levels of EMF or any other energy source are safe, because safety in this area is often a relative concept based on frequency, exposure time, nutritional health, and possible individual sensitivity. Even then, studies are often considered inconclusive plus there is the potential for political and financial agenda to steer perception one way or another.

In order to be fair and equitable while remaining informative, the Scan Tech document has been assembled in order to examine / compare / contrast various safety standards, average environmental levels and references along a continuum to better explain technical measurements in context. This document is offered by Scan Tech to IEIA with permission to post and reproduce in any fashion, crediting Scan Tech as the document creator has been requested by Scan Tech, Dallas, Texas.

Module 9 will address the health and safety overview issues with exposure to various energy sources that have been documented to have a specific health effect as a specific level. The module will bring to the reader a quantitative reference range level as a guideline to assess various energy exposures.

**IEIA - H-SCADA Certification Series**

SCADA systems utilize various types of computers, human sensors, iPods, smart phones, lap tops and many other electronic devices that not only monitor and collect the information under the SCADA system, but may also pose a health hazard risk to the individual using them for long periods of time.

**Assignment**

**LESSON**: Read pages 1 – 9 of the SCAN TECH document as referenced above.

1. The smallest value in a magnetically shielded room is 0.1 nanoGauss. This same value is used to describe a _____.

    A. Squid	B. Repeater	C. Solar Wind	D. Urchin

2. The human heart magnetic field is _____ microGauss.

    A. 10	B. 1	C. 5	D. 0.1

3. A 10 microGauss low level magnetic field is a _____ magnetic field.

    A. Solar wind	B. Galactic	C. Interstellar Space	D. Human Brain

4. Cancer researchers concerned with recent powerline issues are coming up with many reports on oncological effects of very low-level 1 mG ELF _____ fields.

    A. Laser	B. Swamp	C. Electromagnetic	D. Energy

5. _____ or VDTs should produce magnetic fields of no more than 2 mG at a distance of 30 cm (about 1 ft) from the front surface of the monitor and about 50 cm (about 1 ft 8 in) from the sides and back of the monitor.

    A. Hotspots	B. Computer Monitors	C. TVs	D. Radio

6. High magnetic field levels exceeding 100 Gauss (100,000 mG) may cause a temporary visual flickering sensation called _____ which disappears when the field is removed.

    A. Leukemia	C. Magnetophosphenes
    B. Russell-Silver Syndrome	D. Multiple sclerosis

7. The International Commission on Non-Ionizing Radiation Protection (ICNIRP) is an organization of 15,000 scientist from 40 nations who specialize in _____.

   A. Environmental exposures
   B. Radiation protection
   C. Respiratory awareness
   D. UV-A exposures

8. Extreme level magnetic fields are all of the following except, which one of the following:

   A. Surface of a Neutron Star
   B. Strongest Magnetic Spike artificially produced (4-8 microseconds)
   C. Highest Theoretical Magnetic Field
   D. EMF meter

9. The ACGIH Occupational Threshold Limit Values for 60 Hz EMF for the electric field in an Occupational exposure should not exceed for longer than 2 hours at a value of _____.

   A. 25 kV/m    B. 100 V/m    C. 0.1 mg    D. Zero

10. RF Levels and Safety guidelines suggest that humans absorb most radiation between _____ and especially between 77 – 87 MHz.

    A. 30 – 100 MHz    B. SAR 0.4 W/kg    C. 2.6 mW/cm    D. 1 microwatt

11. A typical 802.11b wireless network card will transmit at around 30 mill watts and operates in the _____ frequency band.

    A. 1 microwatt    B. 2.4 GHz    C. 0.09 mW/cm-sq    D. 1 hertz

12. The OSHA 1910 Subpart G 1910.97 Occupational health and environmental control specifically addresses Non-ionizing radiation. It has been ruled _____ for Federal OSHA enforcement and does specific the design of an RF warning sign.

    A. Unenforceable    B. Enforceable    C. Non-contest    D. Unnecessary

# IEIA - H-SCADA Certification Series

13. The unit of a _____ is watts per kilogram (W/kg).

    A. NCRP     B. Scan rule    C. SAR    D. PCS antenna

14. _____ guidelines legally enforceable vary by frequency (10000/frequency^2).

    A. IEEE     B. IEIA     C. EPRI     D. ACS

15. When we talk about radiation effects, we therefore express the radiation as effective dose, in a unit called the _____. This unit also equals 1 Rem.

    A. Gray     B. Beta     C. Sievert     D. Rad

16. Adult workers may receive a whole body dose of _____ per year; minors are restricted to _____ per year. The same value for a minor is restricted to a pregnant woman.

    A. Zero / zero
    B. 1 joule / 1 gray
    C. 5 Rem / 0.5 Rem
    D. 3 neutrons / 4 protons

17. RF fields between 10 MHz and 10 GHz penetrate exposed tissues and produce _____ due to energy absorption in these tissues.

    A. Water     B. Heat     C. Cold     D. Tingling

18. _____ is the maximum actual dose rate received by uranium miners in Australia and Canada.

    A. 3 – 5 mSv/yr    B. 10 mSv/yr    C. 5 Rem    D. 1,000 mSv

19. How many hours may one be exposed to 97 dB Noise level(s) under OSHA Safety Limits.

    A. 0.5 hours    B. 8 hours    C. 3 hours    D. 1 hour

20. The term SQUID stands for Super Quantum Interference Device and is 1,000,000 times less. It is used to measure what type of magnetic field.

    A. Extreme Ultraviolet (EUV)
    B. Low Level Magnetic
    C. High Dose Electromagentic
    D. Nano Tesla Magnetics

## IEIA - H-SCADA Certification Series

### IEIA - H-SCADA Bio-Energy Field Professional Study Guide for Examination

### MODULE 10

**SANDIA Report: SAND 2004 – 5625 Unlimited Release
Printed November 2004
Microfabrication with Fermtosecond Laser Processing
(A) Laser Ablation and Ferrous Alloys,
(B) Direct-with Embedded Optical Wave Guides and Integrated Optics in Bulk Glass**

**TEXT:**

**Sandia Report: SAND2004-5625 Unlimited Release Printed November 2004:** Microfabrication with Femtosecond Laser Processing – (A) Laser Ablation of Ferrous Alloys, (B) Direct-Write Embedded Optical Waveguides and Integrative Optics in Bulk Glass.
Pin Yang, George R. Burns, Jeremy A. Palmer, Marc F. Harris, Karen L. McDaniel, Junpeng Guo, G. Allen Vawter, David R. Tallant, Michelle L. Griffith and Ting Shan Luk, Sandia National Laboratories, Albuquerque, New Mexico, 87185 and Livermore, California

This module will identify femtosecond laser processing that is used to make ferrous alloys with bulk glass. The importance of this lesson is to note that when a custom wave guide is made to attract a signal (frequency) it may not be made of just one compound. The more advanced wave guides are now the size of a few nanometers as a micro bead or diamond dust. The more pure the crystal source the more likely the frequency field may be of a higher range. (Example: UHF vs. Terahertz vs. Microwave, etc.) The bulk glass used in this particular designed wave guide has a ferrous alloy embedded into it, thus giving the glass fibers (nano tubes) an electromagnetic charge that can form piezoelectric materials.

# IEIA - H-SCADA Certification Series

**Assignment**

**LESSON**: Read pages 1 – 78 of the document ~ SANDIA Report: SAND 2004 – 5625 Unlimited Release Printed November 2004 Micro fabrication with Fermtosecond Laser Processing (A) Laser Ablation and Ferrous Alloys, (B) Direct- with Embedded Optical Wave Guides and Integrated Optics in Bulk Glass.

1. Femtosecond laser machining of steel uses simple machine language of _____ and _____ codes.

   A. M and G    B. Cellular and target    C. Dice and mice    D. Hatch and raster

2. When making a laser energy of IMJ at a maximum of 950 mW you would use a _____ laser.

   A. Diamond    B. Pearl    C. Sapphire    D. Coral

3. The _____ program allows for images to be micro machined as meso thunderbirds.

   A. LDRD    B. PETN    C. Lithography    D. Schwartz Schild

4. The glass composition in weight percent, soft point and glass transition temperatures for fused quartz at 99.97 is of all of the following form soft glass composition, except which one does not apply.

   A. $SiO_2$    B. $Na_2O$    C. $B_2O_2$    D. $Al_2O_2$

5. TS, (°C) with a value of 820 % weight glass composition is for _____.

   A. $SF_2$    B. SF57    C. Pyrex    D. CVD Quartz

6. Micro-Raman spectra of laser damaged regions of fuzzed quarts are dependent upon the silica network under a (n) _____.

   A. Extremely high pressure
   B. Vacuum
   C. Temperature
   D. Sub-zero temperatures

# IEIA - H-SCADA Certification Series

7. Light from a 650 nm wave guide and near-field interring distribution of the guide mode illuminates a _____ field with a CCD camera and 0.45 ul pulse energy.

   A. Soccer field    B. 3 D field    C. 2 angle "L" shape    D. Bose effect

8. Laser micromachining is the use of micro channels in the micro fluidic MEM (Micro-electro-mechanical) device for chemical sensing such as _____ on a chip.

   A. DNA Lab    B. Chem Lab    C. Enviro Lab    D. Site Lab

9. Optical absorption difference spectra of femtosecond laser damaged areas can show effect in all of the following layers except which one does not apply.

   A. 2 layers    B. 4 layers    C. "0" layer    D. 3 layers

10. There are specific components that are used in the experimental set-up for photoluminescence measurements by 800 nm femtosecond laser pulses. Identify the component that is not one of the original three.

    A. Glass    B. Spectrometer    C. Plasma    D. Direct write

IEIA - H-SCADA Certification Series

### IEIA - H-SCADA Bio-Energy Field Professional Study Guide for Examination

### MODULE 11

Nano Air Vehicles a Technology Forecast

**Document:**
Nano Air Vehicles A Technology Forecast
by William A. Davis, Major, USAF, April 2007, Blue Horizons Paper, Technology Air War College, Montgomery, Alabama (pgs 1-35)

Additional Information on Urban Nano Air Vehicles, which may contain additional information in the current trends of the integration of nanotechnology into advanced computer systems through SCADA operations:

- **Lockheed Martin Press Release: Lockheed Martin to Design Nano Air Vehicle to Monitor the Urban Battlefield, Posted Wednesday, July 19, 2006**
  http://www.spaceref.com/news/viewpr.html?pid=20389

- **Staninger, Hildegarde.** *"Diagnosing Industrial Environmental Diseases Associated with Exposure to Hazardous Materials vs. Advanced Nano Materials."* Integrative Health Systems®, LLC, Los Angeles, CA © March 6, 2010 (Referenced with written permission from author.)

Module 11 will address the aspects of the specific design for a revolutionary remote-controlled nano air vehicle (NAV) that will collect military intelligence indoors and outdoors on the urban battlefield. These nano devices will have specific chemical rockets or nano payloads for monitoring and a power sensor payload model that will address specific areas of at minimum 1,100 yards or more. The nano machines will transmit images back to a small, hand-held display or other remote source. These small devices as the technology matures may be placed on autopilot that would be able to provide limited or interface with unlimited autonomous operations. The devices will be controlled by an advanced SCADA system as applied to military, private industry developers and government.

The new developments in protecting our homeland from invasion has taken on many different aspects of nanotechnology and will be ever growing in the near future as advancements in the integration of various technologies develop from even earth to space operations for a global security network. During this time it is very important for one to be familiar with the basic applications of hazardous materials and their toxicity vs. nano materials and their toxicity. This aspect is so important to the First Responder during SCADA operations and data collection because the chemical characteristics of a standard hazardous material will not apply to a nano advanced material, where traditionally the element gold is yellow and at a nano size level gold is red.

**Assignment**

**LESSON:** Read the above referenced documents as they apply to the following questions.

1. The future technology of interest to the US Military was the potential development and use of a class of unmanned aerial vehicles (UAVs) called _____ or _____.

   A. F-16 or Skunk Works  C. Black widow or VENOM
   B. Nano-air vehicles or NAVs  D. Compact air machines or ANTS

2. The _____ is known as our nation's technological fighting force and has come to rely on this technology superiority to win our nation's wars.

   A. US Army   B. US Navy   C. US Air Force   D. US Marines

3. _____ defines the NAV "as airborne vehicles no larger than 7.5 cm in length, width, or height, capable of performing a useful military mission at an affordable cost and gross takeoff weight (GTOW) of less than or equal to 10 grams.

   A. CIA   B. DARPA   C. DOJ   D. DHS

4. Micro –UAV (MAVs) have a wing span of _____.

   A. 10 yards   B. 1 inch   C. 6 inches   D. 2 feet

# IEIA - H-SCADA Certification Series

5. A successful operation of the Aero Vironment _____ MAV as part of a DARPA Small Business and innovative Research program allowed the device to fly 30 minutes in both indoor and outdoor environments.

   A. Flash     B. Iron Man     C. Green Star     D. Black Widow

6. All of the following requirements are needed for NAVs, except which one does not apply.

   A. Low Reynolds number airfoils
   B. No guidance system
   C. Energy storage systems
   D. Efficient propulsion

7. The NAV _____ makes the aerodynamic challenge even greater than that overcome by other devices.

   A. Structure     B. Mass     C. Color     D. Speed

8. NAVs under guidance, navigational, sensors and communications are required to operate autonomously and in "_____." This will require much larger processing and sensory requirements.

   A. Trails     B. Rows     C. Swarms     D. Packs

9. The predominant mission of UAVs today is primarily the following:

   A. Endurance, freedom and advancement
   B. Security, tack and control
   C. Intelligence, surveillance and reconnaissance
   D. Travel, sightings and knowledge

10. ISR missions utilizing NAVs are performed within the following areas that are only accessible by man.

    A. Buildings, tunnels, caves and other formations
    B. Sky, homes, vehicles and airports
    C. Food, stores, parks and homes
    D. Foreign boats, cars, radios and towers

# IEIA - H-SCADA Certification Series

11. The following nanotechnology advanced materials may be loaded into a NAV.

    A. Swarms, contaminant, diseases
    B. Nuclear, biological and chemical (NBC) sensors
    C. Manned systems
    D. Launched from soldiers on the ground

12. A nano detector (ND) is purposely built to detect specific molecules or even protein markers from _____.

    A. Biological agents
    B. Chemical agents
    C. Unknown substances
    D. Nano-CMOS

13. NAVs combined with swarming operations could provide some ability in _____ missions.

    A. Nuclear, Biological, and Chemical (NBC)
    B. Flexibility, IRS missions and UAV
    C. Offensive Counter Air (OCA)
    D. Black Widow and VENOM operations

14. Close Air Support (CAS) utilizing NAVs could be used in _____ operations by continually screening the area around a ground force and when an enemy is identified they can quickly attack and neutralize the targets.

    A. IADs sensors-knockouts
    B. Hunter-killer
    C. Multiplier-swarm
    D. Explosives-sensors

15. The swarming NAVs would give the _____ both precise and effective firepower that they can apply at will under Close Air Support (CAS).

    A. Ground commander
    B. Supervisor
    C. Teacher
    D. Law enforcement

16. When a NAV is used in attacking specific targets the best strategic attack is to attack the targets _____.

   A. Home based command
   B. Water supply tanks
   C. Center of gravity
   D. Computer terminals

17. The use of swarms of NAVs could be used as _____ around specific units. They also contain a sensor suite to determine if the enemy is encroaching on the unit's position.

   A. Mobile mine field
   B. Water and air cleaner
   C. Sensor mass brain
   D. Swarm cupcake payload

18. The _____ method is a systematic way to obtain opinions from a panel of experts and obtain consensus without group discussion.

   A. Mohawk     B. Olympus     C. Delphi     D. Roma

19. A Delphi study participant demographic approach may consist on all the following types of participants.

   A. Government, industry, academia
   B. Military, schools, public
   C. Research, engineering, design
   D. Academia and military

20. NAVs require advanced miniature and rugged sensors. Some of the type of sensors that would apply are listed below. Select the one that does not apply.

   A. 3-D scanning laser radar
   B. Micro-Electro-Mechanical Systems (MEMS0
   C. Biologically inspired vision based sensing
   D. Insect-based integrated systems

21. It is truly autonomous that swarming NAVs will require fundamental research into collective behaviors and data structures along with controls. These algorithms will require large amounts of _____ power.

   A. Computational   B. Systems   C. Communication   D. Network

22. Difficulties to be overcome to make swarming NAVs a reality is guidance, navigation, control and _____.

   A. Software   B. Communications   C. Data   D. Networks

23. Progress in _____ devices and systems are required to allow the development of advanced navigation, environment sensing and vehicle control systems.

   A. Sensory   B. Vision   C. Guidance   D. Nano

24. To execute _____ operations the Delphi participants indicated the small size of a NAV payload of approximately 2 grams limits the vehicle's lethality.

   A. Superfund   B. Vehicle   C. Kinetic   D. Space

25. The use of _____ fuel cells for the storage and transfer of power may hold promise for the future size of the NAV.

   A. Sea   B. Solar   C. Biological   D. Air

26. The basic principle in a device that realizes the conversion of biochemical energy into electrical energy is that the process of substrate oxidation by microorganisms or enzymes in the fuel cells offers _____ for electricity production.

   A. Protons   B. Atoms   C. Neutrons   D. Electrons

27. A biological fuel cell uses _____ or other microorganisms for the catalyst in the reaction as compared to chemical fuel cells.

   A. Enzymes   B. Cells   C. Protein   D. Mitochondria

28. The _____ program is the first step for the US Air Force, as they push the envelope of current technology to design, build an demonstrate NAVs for ISR missions.

    A. Joint Capabilities and Integration Development System
    B. US Navy and Private Industrial Applications
    C. DARPA NAV
    D. MAVs and NAVs

29. NAVs may be used in _____ as stated in the Lockheed Martin Press Release of 2006.

    A. Joint Investigations
    B. Time Systems
    C. Urban Battlefield
    D. Public Sectors

30. The diagnosis of industrial environmental disease for hazardous materials will be different for exposure to _____.

    A. Computer networks and chemicals
    B. Advanced nano materials
    C. Swarm NAVs
    D. Sensors and chemical agents

# IEIA - H-SCADA Certification Series

## IEIA - H-SCADA Bio-Energy Field Professional Study Guide for Examination

### MODULE 12

**Technical Supplementary Information and Articles**

**ARTICLES**:

*Extremely Scaled Silicon Nano-CMOS Devices*
Leland Chang, Yang-Kyu Choi, Daewon Ha, Pushkar Ranade, Shiying Xiong, Jeffrey Boker, Chenming Hu and Tsu-Jae King
Proceedings of the IEEE, Vol. 91, No. 11, November 2003

*Designs for Ultra-Tiny, Special-Purpose Nanoelectronic Circuits*
Shamik Das, Alexander J. Gates, Hassen A. Abdu, Garrett S. Rose, Carl A. Picconatto and James C. Ellenbogen IEEE Transactions on Circuits and Systems – I: Regular Papers, Vol. 54, No. 11, November 2007

*General Recipe for Flatbands in Photonic Crystal Waveguides*
Omer Khayam and Henri Benisty
17 August 2009/Vol. 17, No. 17/ Optics Express 14634

This module is specifically designed to address new and relative technical articles that may be used in the design of nano biosensors, electronic noses and other similar technology that will be interfaced into various special niche SCADA systems. The goal of the exercises in this section is to allow the individual who is utilizing this Study Guide to become very familiar with the technology and its advancement into the commercial networks of the world. It is strongly advised that you, as an individual, periodically look at the reference library section of the International Environmental Intelligence Agency (IEIA) or other organizational/governmental libraries that may contain current information in this ever growing technology of advanced materials, computer systems, software and human machine through SCADA.

A collection of articles are cited that will go with the specific questions of the Study Guide for this particular Module.

## Assignment

**LESSON:** Read the specific articles listed in the Study Guide section for Module 12. Please note this area may be updated semi-annually or annually.

1. Ultra tiny, special-purpose nano electronic circuits are the optimal electronic sensor and _____.

    A. Receiver    B. Repeater    C. Actuator    D. Stimulator

2. The ultra tiny sensors have a ____ gait.

    A. Crooked walk    B. Tripod    C. Zebra fish    d. Nano robot

3. Quantum dots and nano wires –based optical electronics may be demonstrated through _____.

    A. Ultra tiny special purpose nano electronic circuits
    B. Nano robots
    C. HP Systems
    D. Parasitic resistors

4. An ultra thin nano CMOS is usually made up of _____ on-insulator.

    A. Silicon    B. Carbon    C. Smart dust    D. Silver

5. Integrated circuits (ICS) are based on _____, which are now called Nano-CMOS devices.

    A. Silicon glass
    B. Silicon implant
    C. Silicon wafer
    D. Silicon MOSFETS

6. Flatbeds (flat wave guides) are usually composed of _____ wave guides.

    A. Photonic crystal    C. Liquid nails
    B. Liquid glass        D. Corn adhesive

# IEIA - H-SCADA Certification Series

7. A 256 Pixel magnetoresistive biosensor microarray may be found in a 0.18 um
   _____.

   A. MITRI     B. CMOS     C. DNA microscan     D. Viral vector

8. Implantable telemetry platform systems of in vivo monitoring within a body may be used for what type of parameters?

   A. Sound     B. Physiological     C. Current     D. Microcontrol

9. The use of implantable telemetry platform systems that may be monitored by SCADA systems are primarily used for _____ and _____, which addresses the quality of patients' care and are aimed at reducing pain and discomfort.

   A. Therapeutic and diagnostics
   B. Sound and light
   C. Speed and temperature
   D. Fabrication and knowledge

10. Viral structure and mechanics are utilized in the applications of using viral proteins in many biosensor small molecular technologies. Viruses and the macromolecular protein shells make up polyhedral shaped viruses. The macromolecular protein is known as a _____. This material is used in advanced materials to create the shape of various sensors.

    A. Capsid
    B. Caspar-King model
    C. Tensergrity
    D. Nano particle

## Appendix A – Acronyms
(Acronyms Utilized under SCADA Applications)

| | |
|---|---|
| ADSS | All Dielectric Self Supporting |
| ADSU | Application Data Service Unit |
| ANSI | American National Standards Institute |
| ANSI-HSSP | American National Standards Institute – Homeland Security Standards Panel |
| ARP | Address Resolution Protocol |
| ASDU | Application Service Data Units |
| | |
| B-ISDN | Broadband Integrated Services Digital Network |
| | |
| CASM | Common Application Service Models |
| CCITT | International Telegraph and Telephone Consultative Committee |
| CDMA | Code Division Multiple Access |
| CMS | Central Monitoring Station |
| COTS | Commercial Off the Shelf |
| CPU | Central Processing Unit |
| CRC | Cycling Redundancy Check |
| | |
| DHS | Department of Homeland Security |
| DMS | Digital Multiplex System |
| DNP | Distributed Network Protocol |
| DNP3 | Distributed Network Protocol Version 3 |
| DoS | Denial of Service |
| | |
| E.O | Executive Order |
| EIA | Electronic Industries Association |
| EOP | Executive Office of the President |
| EPA | Enhanced Performance Architecture |
| EPRI | Electrical Power Research Institute |
| ESD | Emergency Shut Down |
| | |
| FCC | Federal Communications Commission |
| FDMA | Frequency Division Multiple Access |
| FT | Fixed Radio Terminal |

| | |
|---|---|
| GHz | Gigahertz |
| GOMSFE | Generic Object Models for Substation and Feeder Equipment |
| HF | High Frequency |
| HMI | Human Machine Interface |
| HSSP | Homeland Security Standards Panel |
| HTTP | Hyper Text Transfer Protocol |
| I/O | Input/Output |
| IEC | International Electrotechnical Commission |
| IED | Intelligent Electronic Devices |
| IEEE | Institute of Electrical and Electronics Engineers |
| IP | Internet Protocol |
| ISDN | Integrated Services Digital Network |
| ISO | International Organization for Standardization |
| IT | Information Technology |
| ITU-T | ITU Telecommunications |
| LAN | Local Area Network |
| LED | Light Emitting Diodes |
| MAC | Medium Access Control |
| MARS | Multiple Address Radio Systems |
| MHz | Megahertz |
| MMI | Man Machine Interface |
| MTBF | Mean time Between Failure |
| NCS | National Communications System |
| NE/EP | National Security/emergency Preparedness |
| NS | National Security |
| NS/EP | National Security and Emergency Preparedness |
| NT | Network Termination |
| ODBC | Object Oriented Database Connectivity |
| OMNCS | Office of the Manager, NCS |
| OPGW | Optical Power Ground Wire |
| OS | Operating System |
| OSI | Open Systems Interconnection |
| PBX | Private Branch Exchange |
| PCL | Power Line Communication |

| | |
|---|---|
| PCS | Personal Communications Service |
| PI | Program Interruption |
| PLC | Programmable Logic Controller |
| PN | Public Network |
| PSN | Public Switched Network |
| PTM | Point to Multipoint |
| PTP | Point to Point |
| PVC | Polyvinyl Chloride |
| | |
| RF | Radio Frequency |
| RS | Radio Shack |
| RTU | Remote Terminal Unit |
| | |
| RTU/IED | Remote Terminal Unit/Intelligent Electronic Devices |
| | |
| SCADA | Supervisory Control and Data Acquisition |
| SMS | Short Message Service |
| SMTP | Simple Mail Transfer Protocol |
| SQL | Structured Query Language |
| SSB | Single Side Band |
| SSID | Service Set Identifier |
| | |
| TA | Technical Assembly |
| TCP/IP | Transmission Control Protocol/Internet Protocol |
| TDMA | Time Division Multiple Access |
| TETRA | Trans European Trucked Radio |
| TIB | Technical Information Bulletin |
| | |
| UART | Universal Asynchronous Receiver Transmitters |
| UCA | Utility Communications Architecture |
| UHF | Ultra High Frequency |
| | |
| VHF | Very High Frequency |
| VSAT | Very Small Aperture Terminal |
| WAN | Wide Area Network |
| WEP | Wired Equivalent Protocol |
| WOC | Wrapped Optical Cable |

IEIA - H-SCADA Certification Series

## APPENDIX B
## TEXTS and ARTICLES with hot-links for eBook users

## IEIA - H-SCADA Bio-Energy Field Professional
## Study Guide for Examination

### Study Guide Overview

**TEXTS:**

**1 - NCS TIB 04-1: National Communications System**
Technical Information Bulletin 04-1
Supervisory Control and Data Acquisition (SCADA) Systems
Office of the Manager, National Communications System, P.O. Box 4052
Arlington, VA 22204-4052 © October 2004
https://community.emc.com/docs/DOC-17780
http://www.ncs.gov/library/tech_bulletins/2004/tib_04-1.pdf

**2 - Critical Infrastructure: Homeland Security and Emergency Preparedness** by Robert Radvanovsky, ISBN: 0-84593-7398-0
CRC Press/Taylor & Francis (A CRC Press Book), Boca Raton, FL © 2006
http://www.amazon.com/dp/0849373980/ref=rdr_ext_tmb

**3 - Advances in Computers: Volume 71 Nanotechnology**
Edited by: Marvin V. Zelkowitz, ISBN: 978-0-12-373746-5
Elsevier/Academic Press, New York, New York © 2007
Advances in Computers, Volume 71: Nanotechnology: Marvin Zelkowitz
Ph.D. MS BS.: 9780123737465: Amazon.com: Books

**4 - Design for Manufacturability and Yield for Nano-Scale CMOS**
Charles C. Chiang and Jamil Kawa, ISBN: 978-1-4020-5187-6
Springer, Netherlands © 2007
Design for Manufacturability and Yield for Nano-Scale CMOS (Integrated
Circuits and Systems): Charles Chiang, Jamil Kawa: 97814

**5 - FCC Online Table of Frequency Allocations (47 C.F.R.Statue 2.106)**
Federal Communications Commission Office of Engineering and Technology
Policy and Rules Division, Revised April 16, 2013

US Governmental Printing Office, Washington, DC
http://transition.fcc.gov/oet/spectrum/table/fcctable.pdf

## 6 - FDA/OSEL Annual Report 2007

Food and Drug Administration's Center for Devices and Radiological Health (CDRH) and Office of Science and Engineering Laboratories (OSEL)
Larry G. Kessler, Sc.D., Director OSEL/DCRH & Chair, Global Harmonization Task Force

CDRH Reports > FY 2007 OSEL Annual Report
http://www.fda.gov/AboutFDA/CentersOffices/OfficeofMedicalProductsandTobacco/CDRH/CDRHReports/ucm126674.htm

http://www.fda.gov/downloads/AboutFDA/CentersOffices/OfficeofMedicalProductsandTobacco/CDRH/CDRHReports/ucm126717.pdf

## 7 - Guidelines on the Evaluation of Vector Network Analyzers (VNA)

EURAMET (European Association of National Metrology Institutes)
EURAMET cg-2, Version 2.0 (03/2011) Previously EA-10/12 Calibration Guide, Braunschweig, Germany e-mail: secretariat@euramet.org
Phone: +49 531 592 1960

https://www.euramet.org/fileadmin/docs/Publications/calguides/EURAMET_cg-12__v_2.0_Guidelines_on_Evaluation_01.pdf

## 8 - Telemedicine and e-Health: Abstracts from the American Telemedicine Association

Eighteenth Annual International Meeting and Exposition, May 5-7, 2013
Austin, Texas  ISSN 1530-5627  Mary Ann Liebert, Inc. Publishers
*Abstracts from The American Telemedicine Association Eighteenth Annual International Meeting and Exposition May 5–7, 2013—Austin, Texas*
http://www.americantelemed.org/docs/default-source/annual-meeting-2013/presentationabstracts.pdf

## 9 - Section 8: Health Effects Associated with Exposure of Industrial Workers to Radiofrequency Waves

RF Toolkit-BCCDC/NCCEH pages 191 – 216, CDC, Atlanta, GA
http://electromagnetichealth.org/wp-content/uploads/2013/07/RadiofrequencyToolkit_v4_06132013.pdf

# IEIA - H-SCADA Certification Series

**10 - Sandia Report: SAND2004-5625 Unlimited Release Printed November 2004:** Microfabrication with Femtosecond Laser Processing (A) Laser Ablation of Ferrous Alloys, (B) Direct-Write Embedded Optical Waveguides and Integrative Optics in Bulk Glass.
Pin Yang, George R. Burns, Jeremy A. Palmer, Marc F. Harris, Karen L. McDaniel, Junpeng Guo, G. Allen Vawter, David R. Tallant, Michelle L. Griffith and Ting Shan Luk, Sandia National Laboratories, Albuquerque, New Mexico, 87185 and Livermore, California
http://prod.sandia.gov/techlib/access-control.cgi/2004/045625.pdf

**11 - Nano Air Vehicles  A Technology Forecast**
William A. Davis, Major, USAF, April 2007
Blue Horizons Paper, Center for Strategy and Technology, Air War College.  In accordance with Air Force Instruction 51-303, it is not copyrighted, but is the property of the United states government.
http://www.au.af.mil/au/awc/awcgate/cst/bh_davis.pdf

**Staninger, Hildegarde.  "*Diagnosing Industrial Environmental Diseases Associated with Exposure to Hazardous Materials vs. Advanced Nano Materials.*"  Integrative Health Systems®, LLC, Los Angeles, CA © March 6, 2010 (Referenced with written permission from author.)**
http://www.1cellonelight.com/pdf/FinalDiagnosiswithDiagrams.pdf
**PowerPoint Presentation:**
http://www.1cellonelight.com/pdf/DIAGNOSING_ENVIRONMENTAL_DISEASES_vs_ADVANCED_NANO.pdf

**12A - *Extremely Scaled Silicon Nano-CMOS Devices***
Leland Chang, Yang-Kyu Choi, Daewon Ha, Pushkar  Ranade, Shiying Xiong, Jeffrey Boker, Chenming Hu and Tsu-Jae King
Proceedings of the IEEE, Vol. 91, No. 11, November 2003
**IEEE Xplore Abstract - Extremely scaled silicon nano-CMOS devices**
http://ieeexplore.ieee.org/xpl/articleDetails.jsp?arnumber=1240075

**12B - *Designs for Ultra-Tiny, Special-Purpose Nanoelectronic Circuits***
Shamik Das, Alexander J. Gates, Hassen A. Abdu, Garrett S. Rose, Carl A. Picconatto and James C. Ellenbogen IEEE Transactions on Circuits and Systems – I:  Regular Papers, Vol. 54, No. 11, November 2007
http://ieeexplore.ieee.org/xpl/abstractAuthors.jsp?arnumber=4383238

**12C-** *General Recipe for Flatbands in Photonic Crystal Waveguides*
Omer Khayam and Henri Benisty
17 August 2009/Vol. 17, No. 17/ Optics Express 14634
http://www.opticsinfobase.org/oe/fulltext.cfm?uri=oe-17-17-14634&id=184337

http://hal-iogs.archives-ouvertes.fr/docs/00/56/70/27/PDF/00567027.pdf

~~~~~~~~~~~~

# IEIA - H-SCADA Certification Series

## WHAT IS AN

### H-SCADA

Bio-Energy Field Professional? It will Expand Your Career Goals And Horizons

The Reasons and Benefits of

### H-SCADA

Bio-Energy Field Professional Certification

*IEIA is a State of Florida Non-Profit Corporation*

11. SCADA Standards Organizations:
    The Institute Of Electrical and Electronics Engineers (IEEE); American National Standards Institute; Electric Power Research Institute (EPRI), International Electrotechnical Commission and DNP3 Users Group

12. SCADA Governmental and International Interagency Organizations National Communications Systems and Federal Telecommunications Standards Program Academia, Military, Agency and Industrial Overlapping

13. Observations and Conclusions when H-SCADA Assets are in an Asset Investment Program

14. Current and Future Recommendations

*International Environmental Interagency Agency, Inc*
*1770 Algonquin Trail, Maitland, FL 32751*

Image Taken from:
Patent No.: US 6,506,148 B2
Date of Patent: Jan. 14, 2003

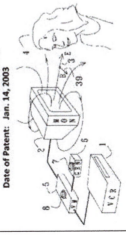

# IEIA - H-SCADA Certification Series

## WHAT IS AN H-SCADA Bio-Energy Professional?

IEIA certifications program are recognized by many international, federal, state and local governmental agencies, the military and industry as Identifying individuals with the capability, education and work experience to do the job right. IEIA is a legally recognized certification and accreditation organization, which has as its mission to provide legal and professional recognition of individuals possessing education, training and experience environmental managers, engineers, technologies, scientists and technicians. Since 2001 to present date, IEIA has provided the skilled professional a professionally recognized credential as one would advance in their career and receive appropriate recognition for their highly revered qualification as they become tapped as the "The Elite" of their profession.

IEIA has been a leader in offering highly professional certification programs through experience, specialty verification, examination and mentorship that has given the greatest personal and professional benefits to our certified professionals.

Our new certification for H-SCADA Bio-Energy Field Professional is in the NEW innovation arena of nano and biotechnology that encompasses our environment, health, medicine, food and safety. It is time to look at the interfacing of our Environment to Human to Machine as professionals move into these new areas of engineering, science and medicine through the integration of technology as well as its use in the field of precision medicine.

The term SCADA (Supervisory Control and Data Acquisition) was originally a type of Industrial Control System (ICS) that has now branched out into medicine, pharmaceuticals, agriculture, academia, military and for many other remote controlled computer systems of data storage, analysis and monitoring. The aspect of Industrial Control Systems is computer controlled systems that monitor and control industrial processes that exist in the physical world. SCADA systems historically distinguish themselves from other ICS systems by being large scale processes that can include multiple sites, and long distances (Example: Earth, Satellite, Space and Cosmos). These processes include industrial, infrastructure and facility-based processes as described below:

Industrial processes include those of manufacturing, production, power generation, fabrication, and refining, and many run in continuous, batch, repetitive, or discrete modes.

Infrastructure processes may be public or private, and include water treatment and distribution, wastewater collection and treatment, oil and gas pipelines, electrical power transmission and distribution, wind farms, civil defense siren systems, and large communication systems.

Facility processes occur both in public facilities and private ones, including buildings, airports, hospitals, schools, ships and space stations. They monitor and control heating, ventilation, and air conditioning systems (HVAC), access and energy consumption. In medicine, one may monitor data gathered from a human body remotely. Recent developments in nanotechnology, biomedicine and life sciences have now incorporated Nano-CMOS, Moffett, MITRI Advanced Computer Systems for the human biological and physiological system into an integrated advanced materials, software and data storage/monitoring.

An Outline of Topics to be covered in the IEIA Study Guide for H-SCADA Bio-Energy Field Professional Certification and Examination will consist of the following:

1. History of SCADA
2. Common System Components
3. Systems Concepts (Nano, Micro and WI FI)
4. Human-Machine Interface
5. Hardware Solutions (Supervisory Station, Operational Philosophy, and Communication Infrastructure and Methods)
6. SCADA architectures: First generation (Monolithic); Second generation (Distributed), Third generation (Networked) and Fourth generation (Cloud plus)
7. Security and Public Privacy Issues
8. Overview of H-SCADA Methodology/Protocols
9. Deploying SCADA Systems: twisted-pair metallic cable, coaxial metallic cable, fiber optic cable, power line carrier, satellites, leased telephone lines, very high frequency radio (terahertz); Ultra High Frequency radio (point to point, multiple address radio systems, spread spectrum radio, microwave radio, and terahertz pulse)
10. Security Vulnerability of SCADA Systems (attacks against SCADA Systems and developing a SCADA Security Strategy)

# IEIA - H-SCADA Certification Series

## ENCOUNTERING CLIENTS CLAIMING EXTERNALLY CONTROLLED "IMPLANTS" ARE AFFECTING THEIR HEALTH: An advisory for healthcare professionals.

Ben Colodzin, Ph.D.

June 2014

Over the last several years I have consulted with individuals from 5 U.S. states and 4 sovereign nations who have claimed to be "implanted" without their consent with some type of technology that affects their physical, mental, and emotional functioning.

I am a therapist and educator in California. Earlier in my career, like most of my colleagues, I had dismissed claims of these types of harassment as an impossibility, and likely a sign of mental illness. One of the first credible accounts that opened my mind was a paper written by the former Chief Medical Officer of Finland, Rauni-Leena Luukanen-Kilde, M. D., titled MICROCHIP IMPLANTS, MIND CONTROL, AND CYBERNETICS, December 2000. Once I became at least somewhat better informed about the very real possible uses of state of the art remote sensing and influencing technologies, I realized that my earlier reflex to simply assume that people who reported such phenomena were "crazy" was based more upon my ignorance than upon scientific fact.

In the 1980's I worked with many Vietnam veterans with Post Traumatic Stress Disorder (PTSD). At that time, PTSD was a newly recognized mental disorder, newly included in the most recent Diagnostic and Statistical Manual of Mental Disorders, 3$^{rd}$ edition. I met many veterans who qualified for the PTSD diagnosis, and many of them had histories of being mis-diagnosed with a wide variety of other mental disorders. Within the established health care bureaucracies, there was a lot of resistance to supplementing their diagnostic evaluative procedures in whatever ways might help them notice if a possible PTSD diagnosis had been overlooked, or if mis-diagnosis had occurred. As people with PTSD increasingly came forward and told their stories, and as our culture increasingly began to listen to them, collectively we got better at noticing when PTSD was present. But before PTSD was officially recognized—and for quite a time after—there was a tremendous amount of mis-diagnosis, and many people were harmed by the failure of health care providers to adequately recognize what was happening for these individuals.

Now in 2014, I see a very similar problem occurring for individuals with advanced implanted technologies non-consensually placed in their bodies. Whereas once we could only speculate if this was real or fiction, we now have toxicological tests that can confirm the presence of advanced nano materials (extremely small manufactured items that can self-assemble into operational machines) in the body, as well as highly sensitive scanning equipment that can detect frequency signals emanating from and directed to a person's body. This means we can actually now use validated tests to discern if a physical system capable of externally altering a person's functioning is actually present and operating.

Although practically no one yet seems to know about these mostly military-type technologies or their fantastic-sounding abilities, and are unaware of the possibility to test for their presence, nonetheless the sound scientific basis of the testing procedures does exist, and increasingly there are individuals whose tests are showing they are indeed imbedded with advanced nano materials, and they are indeed emitting signals and/or resonating with external signals in ways that are just not natural for a human body. This is a reality in the present that should not be ignored. However, without the knowledge that externally controlled machines within the body can massively affect physical and mental functioning, diagnosticians of all health-related disciplines continue to NOT look for the presence of such possible causative factors. This is the established norm today.

How could these technologies possibly be in widespread use without being widely detected? Dr. Luukanen-Kilde, from the perspective of a former Chief Medical Officer of a sovereign nation (Finland), offered this view in December 2000:

*One reason this technology has remained a state secret is the widespread prestige of the psychiatric DIAGNOSTIC STATISTICAL MANUAL IV, produced by the U.S. American Psychiatric Association (APA) and printed in 18 languages. Psychiatrists working for U.S. intelligence agencies no doubt participated in writing and revising this manual. This psychiatric "bible" covers up the secret development of MC (mind control) technologies by labeling some of their effects as symptoms of paranoid schizophrenia.*

*Victims of mind control experimentation are thus routinely diagnosed, knee-jerk fashion, as mentally ill by doctors who learned the DSM "symptom" list in medical school. Physicians have not been schooled that patients may be telling the truth when they report being targeted against their will or being used as guinea pigs for electronic, chemical and bacteriological forms of psychological warfare.* [END QUOTE]

However, as in the earlier example I mentioned about PTSD, eventually enough evidence can accrue that previously disregarded information reaches a critical mass, and the possibility that something previously undetected is playing a role begins to gain support. That is the point we seem to be at now, regarding scientific evidence that there truly are "targeted individuals" being externally influenced in ways we who are not privy to classified information previously thought impossible.

Now it can be proven that advanced nano machines do exist inside some human bodies, and it can be proven that the knowledge and technical infrastructure exists to transmit signals remotely that powerfully affect the brain and other human functions in selected or "targeted" individuals. The capability is now being developed to track these external controlling frequency signals back to their sources. With this development we are entering an era where accountability for this totally unregulated human experimentation and manipulation may become possible for these previously invisible activities. In this situation and with these technologies now poised to proliferate, the ethical health provider needs to be informed about this sad state of affairs.

I make this declaration to witness to health care professionals around the world: if you have clients who claim they are being targeted or "chipped" and present the symptoms that are commonly associated with this phenomena (see references below), do not automatically assume they are delusional. They may be, but they may not as well. Or, they may have delusions that are not indicative of organic psychosis, but instead of artificially induced, virtual reality-based, externally transmitted delusions. For which there is no diagnostic category in current psychiatric thinking. With the application of these technologies increasing as their sophistication grows, we are in a whole new diagnostic ball game.

And if such clients come to you with Raman spectroscopy tests confirming the presence of advanced nano materials in their body fluids, and sophisticated scanning tests showing unusual frequency emissions emanating from their bodies or other results that are not in the normal range for human beings, it may be time to re-think the possible, to get more informed about the state of the art of brain-altering weapons systems, and to listen to what these people who are calling themselves "targeted individuals" have to say.

Misdiagnosis is the common lot of almost everyone who has become an experimental test subject for these technologies. The lack of recognition of what is actually happening for them causes perhaps as much harm as the aggressive technological attacks they are subjected to. For all health practitioners who find the injunction to DO NO HARM a worthy standard, be prepared to investigate the possibility that "targeted individuals" claims of being externally programmed may be real, prior to reaching other diagnostic conclusions.

Ben Colodzin, Ph.D.

June 21, 2014

# IEIA - H-SCADA Certification Series

# IEIA - H-SCADA Certification Series

## H-SCADA PROTOCOL EVALUATION METHODS and EQUIPMENT

The evaluation and assessment protocols used in the H-SCADA methodologies as adapted by IEIA for certification/investigations are based on the original developmental work of Melinda Kidder, BS, PI, CESCO, Columbia Investigations, Columbia, MO.

The equipment used during the evaluation and assessment protocols are listed per the appropriate methods as referenced below:

**Method Standard One: EMF Field Meter**

Purpose: Detection of electromagnetic radiation from client.

- Measures electromagnetic field radiation
- LCD Display of EMF level in milliGauss or microTesla (documented in microTesla)
- Provides accurate measurements to 4% over a measuring range of 0.1 to 199.1 mGauss (0.01 to 19.99µTesla)
- ELF Frequency bandwidth of 30 to 300 Hz
- Single axis – sampling 2.5 times per second

**Method Standard Two: RF Frequency Detector with Bar Graph**

Purpose: Detection of radio frequency coming to or from client in the specified range.

- Frequency range of 1MHz-3GHz
- Sensitivity: Less than 5 mV
- Microprocessor filtration circuitry allowing squelch adjustment to diminish RF noise
- High sensitivity LCD bar graph
- Used both with and without "rubber duck" antenna during this testing

**Method Standard Three: UHF Bluetooth Wireless RFID Scanner/Reader/Writer**

Purpose: Detection of RFID chips/tags in client within specified range, reading any code written on any chip/tag detected, and optionally re-writing chip/tag code.

- Frequency Range: EU865-868MHz; US902-928MHz
- Standards Supported: EPC Class 1 Gen 2
- Antenna: Detachable, Circularly Polarized with Optional 2D Scanner
- Nominal Read Range: Up to 13'/4m
- Nominal Write Range: Up to 4'/1.22m
- Field: 150 ° Forward Facing (approx.) Measured From Front of Device

**Method Standard Four: UV Light Inspection**

Purpose: Inspection of eyes of client for determining UV fluorescence and noting color.

- UV Light: 395 nm - 405 nm

**Method Standard Five: Night Vision Scope**

Purpose: Detection of bio-glass, infrared glass and/or other nano-glass materials for the inspection of infrared reactivity of client.

- Image capture capability
- Infrared intelligence

**Method Standard Six: Metal Detector**

Purpose: Detection of subcutaneous metals in client.

- Operating Temperatures -35º F (-37º C) to 158º F (70º C)
- Operating Frequency: 95 kHz
- Tuning: Automatic
- Scan Area: 3.5" and 360º plus tip
- Ultra-sensitive response to metal objects up to 4" depth
- Accurate detection of all ferrous, non-ferrous and stainless steel objects

**Method Standard Seven: Portable Spectrum Analyzer**

**Purpose:** Detection of radio frequency coming to or from the client in the specified range for it's appropriate setting.

- **Frequency Range:** 15 – 2700 MHz
- **Sensitivity:** typ. ± 10 ppm
- **Sensitivity Level:** typ. ± 3 dBm
- **Resolution f:** min. Bandbreite/112 typ.
- **Resolution Level:** 0.5 dBm typ.
- **Setting Accuracy f:** 1 kHz
- **Broadband Displayed:** 112 kHz - 600 MHz
- **Antenna Jack:** 2x SMA
- **Antenna Impedance:** 50Ω
- **Display:** LCD w/background light, 128 x 64 px
- **Dynamic Range:** -115 – 0 dBm typ.
- **Noise Floor:** -115 dBm typ.
- **Max Input Level:** +5 dBm
- **Weight:** 185 grams
- **Dimension w/out antennas:** 113 x 70 x 25m

**Method Standard Eight: Thermal Imaging Camera**

**Purpose:** Thermal imaging of client.

- True Thermal Sensor
- 206 x 156 Array
- 32,136 Thermal Pixels
- 12μ  Pixel Pitch
- Vanadium Oxide Microbolometer
- 36 º Field of View
- Magnesium Housing
- Long Wave Infrared 7.2 – 13 Microns
- -40C to 330C Detection
- <9Hz

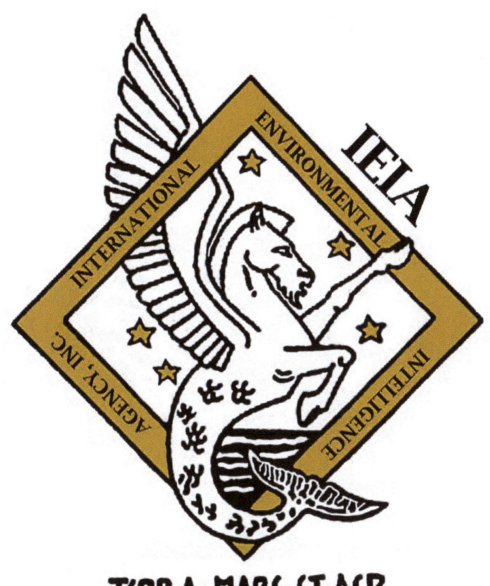

# IEIA - H-SCADA Certification Series

## Answers to Study Guide Questions

### MODULE 1 - Key to Study Guide Questions Lessons 1 through 3

**Lesson 1**

1. D
2. A
3. A
4. D
5. C
6. D
7. B
8. C
9. D
10. B
11. B
12. A
13. C
14. D
15. D
16. B
17. D
18. C
19. A
20. D
21. D
22. B
23. A
24. C
25. A
26. D

**Lesson 2**

1. D
2. C
3. D
4. A
5. C
6. D
7. A
8. D
9. A
10. D
11. B
12. A
13. D
14. B
15. C
16. A
17. D
18. D
19. D
20. A
21. A
22. A
23. C
24. C
25. A
26. C
27. B
28. B
29. D

**Lesson 3**

1. A
2. C
3. A
4. D
5. D
6. C
7. A
8. B
9. D
10. D
11. D
12. B
13. A
14. B
15. C
16. A
17. C
18. B
19. D
20. All Dielectric Self Supporting
21. Broadband Integrated services Digital Network
22. Electric Power Research Institute
23. Federal Communications Commission
24. High Frequency
25. Hyper Text Transfer Protocol
26. Distributed Network Protocol Version 3
27. Cyclic Rudundancy Check
28. Office of the Manger, NCS
29. Radio frequency
30. Remote Terminal Unit
31. Transmission Control Protocol/Internet Protocol

###
**Module 2 - Answers to Study Guide Questions**

**Lesson 1- Chapter 1**
1. B
2. D
3. D
4. C
5. B
6. A
7. C
8. A
9. D
10. C
11. B
12. D
13. A
14. D

**Lesson 3- Chapters 3, 4 & 5**

**Ch. 3**
1. A
2. C
3. C
4. C
5. D
6. A
7. D
8. B
9. A
10. D

**Ch. 4**
1. C
2. D
3. D
4. D
5. D
6. D
7. A
8. C
9. D
10. B
11. A
12. B
13. C
14. D
15. D
16. C
17. D
18. B
19. A
20. C
21. D
22. C

**Ch. 5**
1. B
2. D
3. A
4. D
5. B

**Lesson 2– Chapter 2**
1. A
2. B
3. C
4. C
5. D
6. C
7. A
8. A
9. D
10. C
11. D
12. A
13. B
14. D
15. C
16. B
17. C
18. B

## Lesson 4: Chapters 6, 7 & 8

| Ch. 6 | | Ch. 7 | | Ch. 8 | |
|---|---|---|---|---|---|
| 1. | A | 1. | D | 1. | B |
| 2. | B | 2. | C | 2. | D |
| 3. | A | 3. | D | 3. | D |
| 4. | C | 4. | A | 4. | B |
| 5. | D | 5. | A | 5. | A |
| 6. | B | 6. | D | 6. | D |
| 7. | A | 7. | D | 7. | B |
| 8. | D | 8. | B | 8. | D |
| 9. | A | 9. | C | 9. | C |
| 10. | B | 10. | C | 10. | A |
| 11. | C | 11. | A | 11. | C |
| 12. | C | 12. | D | 12. | A |
| 13. | A | 13. | B | 13. | D |
| 14. | B | 14. | C | 14. | B |
| 15. | D | 15. | B | 15. | A |
| 16. | D | 16. | A | 16. | A |
| 17. | A | 17. | C | 17. | A |
| 18. | B | 18. | D | 18. | C |
| 19. | C | 19. | C | 19. | A |
| 20. | B | 20. | B | 20. | B |
| | | 21. | A | 21. | D |
| | | 22. | D | 22. | D |
| | | | | 23. | A |
| | | | | 24. | C |
| | | | | 25. | C |

## Lesson 5: Chapter 9, 10 & 11

| Ch. 9 | | Ch. 10 | | Ch. 11 | |
|---|---|---|---|---|---|
| 1. | D | 1. | D | 1. | B |
| 2. | D | 2. | A | 2. | C |
| 3. | B | 3. | C | 3. | B |
| 4. | A | 4. | B | 4. | A |
| 5. | C | 5. | B | 5. | C |
| 6. | D | 6. | A | 6. | C |
| 7. | B | 7. | C | 7. | D |
| 8. | A | 8. | D | 8. | B |
| 9. | B | 9. | D | 9. | C |
| 10. | B | 10. | A | 10. | C |
| 11. | C | | | | |
| 12. | D | | | | |

13. B
14. C
15. D
16. B
17. A
18. C
19. A
20. C
21. B
22. A
23. C
24. D
25. A
26. A
27. C

###

## Module 3 - Answers to Study Guide Questions

| **Lesson 1** | **Lesson 2** | **Lesson 3** | **Lesson 4** | **Lesson 5** | **Lesson 6** |
|---|---|---|---|---|---|
| 1. B | 1. A | 1. A | 1. B | 1. B | 1. A |
| 2. A | 2. A | 2. D | 2. D | 2. D | 2. C |
| 3. A | 3. B | 3. A | 3. A | 3. C | 3. B |
| 4. C | 4. A | 4. C | 4. A | 4. C | 4. A |
| 5. C | 5. B | 5. C | 5. D | 5. C | 5. B |
| 6. A | 6. D | 6. D | 6. A | 6. B | 6. B |
| 7. C | 7. C | 7. B | 7. D | 7. A | 7. D |
| 8. D | 8. C | 8. B | 8. C | 8. D | |
| 9. A | 9. C | 9. D | 9. A | 9. C | |
| 10. B | 10. A | 10. B | 10. C | 10. B | |
| 11. C | 11. B | 11. A | 11. A | 11. D | |
| 12. B | 12. D | 12. A | 12. B | | |
| 13. A | 13. B | 13. C | 13. C | | |
| 14. B | 14. A | 14. B | 14. D | | |
| 15. A | 15. A | 15. D | 15. A | | |
| 16. C | 16. B | 16. B | | | |
| 17. D | 17. C | 17. A | | | |
| 18. B | 18. D | 18. C | | | |
| 19. C | 19. A | 19. B | | | |
| 20. C | 20. C | 20. C | | | |
| 21. D | 21. B | 21. C | | | |
| 22. B | 22. A | 22. D | | | |

| | | |
|---|---|---|
| 23. A | 23. B | 23. A |
| 24. A | 24. A | 24. B |
| 25. D | 25. A | 25. B |
| 26. A | 26. C | 26. B |
| 27. C | 27. C | 27. D |
| 28. D | 28. C | 28. A |
| 29. B | 29. D | 29. C |
| 30. A | 30. A | 30. A |
| 31. A | 31. C | |
| 32. D | 32. B | |
| 33. C | 33. A | |
| 34. C | 34. B | |
| 35. A | 35. D | |
| 36. C | 36. D | |
| 37. B | 37. C | |
| 38. D | 38. D | |
| 39. A | 39. C | |
| 40. B | | |
| 41. B | | |
| 42. A | | |
| 43. C | | |
| 44. D | | |

###
## Module 4 - Answers to Study Guide Questions

**Lesson 1**
1. C
2. D
3. A
4. A
5. B
6. D
7. A
8. C
9. B
10. D
11. A
12. A
13. C
14. B
15. D

**Lesson 2**
1. D
2. A
3. A
4. A
5. D
6. B
7. C
8. A
9. D
10. A

16. B
17. A
18. C
19. C
20. D
21. C
22. A
23. A
24. A
25. A
26. B
27. B
28. C
29. D
30. C
31. A
32. B
33. C
34. C
35. A
36. A
37. A
38. B
39. A
###

## Module 5 - Answers to Study Guide Questions

**Lesson 1**
1. A
2. C
3. B
4. D
5. B
6. D
7. A
8. C
9. B
10. A
11. C
12. D

**Lesson 2**
1. A
2. B
3. C
4. D
5. C
6. D
7. A
8. B
9. A
10. B
11. C
12. D

###

**Module 6 - Answers to Study Guide Questions**

1. A
2. C
3. B
4. D
5. C
6. A
7. B
8. D

###

**Module 7 – Answers to Study Guide Questions**

1. B    8. C
2. A    9. B
3. D    10. A
4. D    11. A
5. C    12. A
6. A    13. B
7. D

###

**Module 8 - Answers to Study Guide Questions**
**Lesson 1 and Lesson 2**

1. A    7. B
2. C    8. C
3. B    9. B
4. C    10. D
5. D
6. C

###

**MODULE 9 – Answers to Study Guide Questions**

1. A
2. A
3. B
4. C
5. B

6. C
7. B
8. D
9. A
10. A
11. B
12. A
13. C
14. A
15. C
16. C
17. B
18. A
19. C
20. B

###

## MODULE 10 - Answers to Study Guide Questions
1. A
2. C
3. A
4. B
5. C
6. A
7. B
8. B
9. C
10. C

###

## Module 11 - Answers to Study Guide Questions
1. B
2. C
3. B
4. C
5. D
6. B
7. A
8. C
9. C
16. C
17. A
18. C
19. A
20. D
21. A
22. B
23. A
24. C

10. A     25. C
11. B     26. B
12. A     27. A
13. C     28. C
14. B     29. C
15. A     30. B

###

**MODULE 12 - Answers to Study Guide Questions**

1. A      9. A
2. B     10. A
3. A
4. A
5. D
6. A
7. B
8. B

###

**NOTES:**

Printed in the USA
CPSIA information can be obtained
at www.ICGtesting.com
LVHW080814091023
760516LV00012B/1284